A FORESTER'S LEGACY

The Life of Joseph E. Ibberson

FORESTER, TREE FARMER, PHILANTHROPIST

by HENRY D. GERHOLD

Published by Pennsylvania Forestry Association in association with Stackpole Books

Printed in U.S.A.

10 9 8 7 6 5 4 3 2 1

Library of Congress Cataloging-in-Publication Data

Gerhold, Henry D.
A forester's legacy : the life of Joseph E. Ibberson / Henry D. Gerhold.
p. cm.
Includes bibliographical references and index.
ISBN-13: 978-0-8117-3390-8 (hardcover)
ISBN-10: 0-8117-3390-4 (hardcover)
1. Ibberson, J. E. 2. Foresters–Pennsylvania–Biography. I. Title.
SD129.I22G47 2007
634.9092–dc22
[B] 2006027158

CONTENTS

PREFACE

Norman Lacasse, on behalf of the Pennsylvania Forestry Association, in June 2003 persuaded me to write a biography of Joseph Ibberson. Besides documenting Joe's life and achievements, it would tell the story of various people and events in the history of the Bureau of Forestry. Joe also hoped that philanthropy by others would be encouraged when they learn about his gifts of forestlands and financial contributions, thus expanding the acreage of public parks for the benefit of all Pennsylvanians.

The following spring, I began a series of interviews at Ibberson's home, and these continued until 2006. On many occasions, Charles Strauss, director of Penn State's School of Forest Resources, also participated actively. Strauss and I occasionally met with Ibberson since 2001 in connection with plans for the Joseph E. Ibberson Chair in Forest Resources Management. We had developed mutual respect and trust, so the interviews for the biography proceeded candidly. We traveled together to places where Joe had grown up and worked at various jobs, to

Henry Gerhold interviewing Joe Ibberson in his kitchen.

Joe's good friends Nila Grimes and Charles Strauss helped with the interviews.

take photos and talk with people who had known him. Joe's close friend and confidant Nila Grimes also provided candid comments and observations.

Though in his late eighties, Ibberson has an excellent memory of all kinds of details about every stage of his life. He also retrieved from his files many documents, photographs, and financial records that authenticated and supplemented his recollections. I recorded conversations with Joe and took notes to use in compiling his story. To supplement the interviews, Joe wrote rather extensively about his recollections, adding many details to the oral and written records. Accordingly, some of the paragraphs, even some chapters, are largely autobiographical.

ACKNOWLEDGMENTS

Financial support generously given by the Pennsylvania Department of Conservation and Natural Resources honors Joseph Ibberson as the father of its Conservation Area Program. Ibberson conceived of the program and developed the idea with Roger Fickes, Gregory Schrum, and DCNR Secretary John Oliver, and his gift of the first conservation area inspired other donations.

Members of the Pennsylvania Forestry Association also have contributed funds for the Ibberson biography and have assisted in preparing and reviewing the manuscript, supervising publication arrangements, and publicizing the book. Norman Lacasse and John Laskowski, in particular, devoted hours to help with the selection of illustrations and arrangements for publication.

I am grateful to all of the friends and family of Joseph Ibberson who contributed their thoughts and insights and thus helped make this descriptive portrait more comprehensive and colorful. They include Jean Craighead George, author of popular novels for children; Paul Troutman Jr., Bureau of Forestry service forester; Alexander Boyd and Mead Shaffer, both influenced by Ibberson to donate property; Nila Grimes, close friend and partner in philanthropy; nephew Glenn Ferguson and his mother, Beverly Ferguson; Rick and Betty Petroski and their sons, who worked on Ibberson's tree farms; and logger Ben Mentzer. Betty Petroski also was a great help with proofreading. A number of people who had worked with Joe or knew him in his college days added some interesting observations. Collectively, these men confirmed some of the common experiences they shared with Joe and elaborated on their relationships with him. My sincere thanks to John Bitzer, William Corlett, Alex Day, Thomas Delong, Roger Fickes, William Forrey, Ronald Gale, John C. Good, Norman Lacasse, Patrick Lantz, James Nelson, W. R. "Dick" Rossman, Frank Schrey III, Gregory Schrum, Leroy D. Schaller, Michael Taras, and Jack Winieski.

I am also thankful that I was permitted to use some of my time as an employee of the School of Forest Resources of the Pennsylvania State University to work on this most enjoyable project. Most of all, I am appreciative of the many hours I was able to spend with Joe, getting to know this unique individual, his philosophy, and his great achievements. Our friendship became closer during the course of the book project.

–Henry D. Gerhold

CHAPTER 1
Boyhood Aspirations

What a revelation Joe Ibberson had that day: "I'm gonna be a forester!" For some time, this young boy had been thinking about forestry as a vocation. The idea took root and grew as Joe listened to his father encouraging his eldest son, Henry, to become a forester. Henry had frequent conversations with a Penn State forestry student named Charles Henninger, and Joe would listen in with great interest. He also read the forestry books and magazines that Henninger and Henry brought to the house.

The very thought that he too could become a forester fascinated Joe. The more he contemplated a career in forestry, the more alluring it became. It was a photo in a book that finally clinched the decision—a picture of the Father of Pennsylvania Forestry, Dr. Joseph Trimble Rothrock, traveling through the forest in a horse-drawn buckboard wagon with his camera and lantern slides to lecture all over the state. If a successful medical doctor could make such sacrifices for a cause he deemed so important, Joe was convinced of the necessity for foresters to continue that effort. He decided at that moment that forestry was the only profession for him. Brother Henry was aware of Joe's fascination and teased Joe that he had a tree growing out of his head.

The book Joe was reading that day described Rothrock's unique contributions to forestry. Rothrock had campaigned throughout the state in the late 1800s to convince people that forests were being destroyed by logging, fires, and erosion. He traveled far and wide, often over rough roads, from lecture to lecture. His slides, projected by an oil-burning lantern, showed the devastation of "cut-out-and-get-out" logging practices. More photos followed, illustrating the flames and smoke of fires that often followed the logging, consuming vast acreages of Pennsylvania's forests and exposing the land to erosion. After building public support, Rothrock persuaded the legislature to purchase cutover lands for forest reserves. Through his dedication and perseverance, he led the way in creating Pennsylvania's magnificent state forests and founded the Mont Alto Forest Academy, which trained foresters to tend them.

Inspired by Rothrock's deeds, Joe decided then and there that he would devote his life to forestry, "without a doubt!" Even at the young age of twelve, he realized that much work still was needed to conserve and improve the forests of Pennsylvania for future generations. "Forestry had a magic ring to it," Joe recalls, and it appeared to be a noble profession. Even after Rothrock's death, he had recruited another forester to his noble cause, a young lad who also happened to be named Joseph.

Putting the book aside, Joe reflected on the many happy hours and days he had spent in the woods. Trees just a short walk from his boyhood home always beckoned to him. He often rambled alone or with friends through the forests that

J. T. Rothrock's inspiring career left Joe no doubt that he wanted to be a forester

stretched for miles over several ridges. As soon as he entered the cool, damp shade on a hot summer day, the sights, sounds, and smells enticed him to explore ever deeper. On crisp autumn days, he sometimes was excited by the rustling of dry leaves that told him he had startled a deer. Joe took great pleasure in hunting and fishing with friends and family. No wonder he loved the forest so greatly!

As a boy of six, he already was gathering kindling wood for his mother's cook-stove, transporting the sticks in a little coaster wagon. He always had a penchant for doing something that would be productive and thus give him something to show for his efforts. This innate tendency that surfaced in his youth would be expressed throughout his life, for he continued to derive pleasure from doing things that were productive. And at this formative moment as a teenager, reading about Rothrock, he dreamed about what he could do for the forests that meant so much to him.

Joseph's mother was a source of inspiration of a different kind. She often read from the Bible to the children before bedtime. She had a knack of getting them to feel as though they were personally involved in a story and experiencing it themselves. A favorite of hers was the tale of Joseph and his coat of many colors. The biblical Joseph had progressed from an Egyptian slave to advisor and confidant of the Pharaoh. He saved the people from famine by forecasting seven years of plenty and seven years of crop failure. Joe's mother emphasized how Joseph, after becoming a ruler of the Egyptians, was able to provide for his own family and many other people.

Thinking back years later, Joseph is convinced that his mother had read this story so many times to encourage him to be like his biblical namesake. She pointed out that even the youngest son could accomplish much in life. And he did not disappoint her, as his life's story will reveal. So Joseph, emboldened by his mother's high hopes for him, formulated aspirations that led to a dual mission for his life: help preserve the forests he loved and do good deeds for people. As he matured, his focus shifted to conserving forests for all people to enjoy.

These influences—his mother, the Rothrock book, and happy days in the woods—turned out to be seminal ones in Joseph's life. Most teenagers have not yet thought about a goal in life, nor have they settled on what career they want to pursue. By his teens, however, Joe already had an idealistic vision, had set clear goals, and was determined to pursue them. From then on, he worked hard at various jobs and saved his money so he could study forestry in college.

As the next years passed, he prepared diligently to follow his dream despite obstacles he encountered. Joe found ingenious ways to exploit opportunities that came his way, and sometimes he created the opportunities. His story illustrates how a lad, with some initial parental guidance, can arise from humble beginnings to great achievements. In pursuing his goals, he displayed determination, innate intelligence, hard work, and a knack for building positive relationships with people around him. The tale of his life is the epitome of American success stories (see Appendix A). Little did Joe realize then what a great, positive influence he would have on the profession of forestry, forestry education, and the many people who rely on forests for employment, wood products, pure water, and recreation. He fostered a renaissance in the conservation of Pennsylvania's forests, which in turn enhanced the many environmental, social, and commercial benefits that these forests provide to people.

CHAPTER 2

The Ibberson Family

Joseph Edgar Ibberson was born March 12, 1917, in the small village of Lykens in Dauphin County, Pennsylvania. The youngest of four children, he had two brothers and a sister, who was the oldest child. When Joseph was one year old, his parents bought a house at 524 West Street, a few blocks from the center of Lykens. This was his residence until 1949, when he moved to Harrisburg, where he has resided ever since.

Electricity was installed in the family home when Joe was about five or six years old. Before that, kerosene lamps provided light, and coal and wood were used for heating and cooking. There was a vegetable garden in the backyard, as well as an outhouse. Joseph attended church regularly with his family and had a wholesome family life.

Joseph's father, William Ibberson (1890–1964) was a building contractor by trade. He helped build the four-story Lykens Hotel, the largest building in town, which served as the center of many business and social functions. Although he went only as far as the third grade in school, he was very smart and especially good in math. William later became a coal miner during the Depression, when his building business foundered. He was about six feet tall, and all his sons were six feet or more in height.

Joe's mother, Beulah Zimmerman Ibberson.

Joe's childhood home at 524 West Street, Lykens, Pennsylvania, as it appeared in 2004.

Joseph's mother was born Beulah M. Zimmerman in 1892 near Elizabethville to a Pennsylvania Dutch family and grew up on a farm a few miles from Lykens. William and Beulah were married about 1909. Beulah, who spoke the Dutch dialect, displayed all the good qualities of a Dutch wife and mother and prepared delectable meals. At about five feet, seven inches, she was taller than the average height of women at that time. She was very well organized and managed money wisely. The butcher, milkman, and produce farmers all came to the Ibberson home in Lykens with horse-drawn wagons to peddle their goods. Beulah always bought the best cuts of meat and garden produce from them at bargain prices, and they seemed to enjoy dealing with a shrewd customer like Joseph's mother.

Joe planted the chestnut tree behind his childhood home in 1948, fifty-six years before this photo.

Joe's oldest brother, Henry, was four years older. Nicknamed Hen, he was the tallest of the siblings into adulthood. Hen's father intended for him to become a forester, and he did look into the profession. But at the age of fourteen, Henry had a job as clerk at the corner drugstore and soda fountain and became well known locally, especially among the young ladies, because he was already over six feet tall. So Henry developed interests other than forestry and he did not attend college. He eventually found employment in construction and worked for many years at the Naval Depot in Carlisle.

Brother Ray, two years older than Joe, was happy-go-lucky, personable, and well liked by all. Ray was a very good athlete, and he excelled as a baseball pitcher and a halfback on the Lykens football team. He ran a delivery truck for a bakery and

The home where Joe's mother grew up, several miles from the Eisenhower ancestral home.

was a great salesman. Later he worked as a brakeman for the railroad in the caboose of a coal train that ran from Harrisburg to Renovo. His athletic skill stood him in good stead, for to set the brakes by hand, he had to run from the roof of one railroad car to the next while they were in motion, at the same time watching for overhead obstacles that could knock him off.

As the youngest, Joe wore mostly hand-me-down clothes. His wire hoop toy came from a broken nail keg.

Joseph, front left, with brothers Ray, Henry (in back) and sister Ruth.

Sister Ruth, six years older, was bright and attractive, with blond hair and blue eyes. She graduated from Shippensburg Normal School and became a schoolteacher, first teaching at Millersburg at age sixteen, and later at Haddonfield, New Jersey. She married Harry Stutzman, a doctor's son from Tower City, and they had two children. Ruth later returned to teaching at Millersburg and Lykens. Former students often came to speak with her to express appreciation for her part in their education. Her brothers called her "the general," as it seemed to them that she ran everything.

Joseph, being the youngest of four children, got hand-me-down clothes from his two brothers and even his sister. He had simple toys, such as a hoop from a broken nail keg that he rolled with a stick. The family couldn't afford more expensive toys, as their income was modest, and everyone was expected to be helpful and frugal.

The Eisenhower ancestral home at Elizabethville, Pennsylvania.

Historical plaque at the Eisenhower ancestral home.

Two uncles, Harvey Hoover and Charles Rowe, were farmers who had sons and daughters near Joe's age. Uncle George, his father's oldest brother, was unusually talented and could do practically anything, though he had a stubborn streak. On family camping trips to Beaver Camp, Uncle George served as an excellent cook. He was an experienced coal miner, adept at tunneling through rock, and also served as president of the Lykens borough council.

Joseph's mother and father also shared a common ancestor with General Dwight D. Eisenhower, who became president of the United States, and his brother Milton, who served as president of both Penn State and Johns Hopkins universities. All are direct descendants of John Matter, who was born in 1732 in Alsace Lorraine and emigrated to America in 1751. Whether or not this conferred a genetic advantage in leadership qualities, Joseph Ibberson at least drew inspiration from the familial relationship.

Joseph Ibberson might have been destined to become a forester by genealogy. His grandfather William Ibberson was born April 4, 1859, at Huddersfield, England, east of Manchester. William came to America in 1879 and married Louisa Jane Stuppy in 1883. A genealogist traced the Ibberson name to a Danish general who had invaded England in one of the Danish invasions during the ninth and tenth centuries. In 1016, the Danish king was made king of England. General Ibberson was given a large tract of land in the county of Huntingdonshire for his exploits in battle. He had a large estate with lovely buildings, and some of Joe's

Genealogy of the Ibberson and Eisenhower Families, Showing that the Eisenhowers and Both Parents of Joseph E. Ibberson Had a Common Ancestor

John Matter (1732–1802)
Ann Barbra Arnold

John Matter Jr. (1759–1832)
Elizabeth Berger (1762–1832)

Michael Matter (1763–1852)
Anna Marie Romberger (1771–1838)

Jacob Matter (1790–1842)
Sara Fisher (1789–1832)

Solomon Matter (1806–1877) Isobelle Rowe (1810–?)	Henry Matter (1796–1868) Ann Mary Deitrich (1803–1865)	Joseph Matter (1814–1857) Ana Mary Yerges (1815–1893)
Mary Ann Matter (1832–1910) Amous Cooper (1824–1894)	Rebecca Matter (1825–1890) Jacob F. Eisenhower (1826–1906)	Anna Maria Matter (1836–1890) Ludwig Stuppy (1835–1887)
Julia Ann Cooper (1868–1933) Harry T. Zimmerman (1863–1905)	David Jacob Eisenhower (1863–1942) Ida Elizabeth Stover (1862–1946)	Louisa Jane Stuppy (1862–1935) William Ibberson (1859–1899)

Beulah M. Zimmerman (1892–1957) **William Ibberson II** (1890–1964)

← *cousins* → **Dwight D. Eisenhower** (1890–1969) ← *cousins* →
Mamie Geneva Doud (1896–1979)

and

Milton S. Eisenhower (1899–1985)
Helen E. Eakins (1904–1955)

Joseph E. Ibberson (1917–)

Joseph's great-grandparents, Ludwig and Anna Maria Stuppy, are seen at front center. Their daughter Louisa Jane, at upper left, married Joseph's grandfather William Ibberson.

relatives in Minnesota have pictures of them. The genealogist also found records of a John Ibberson at Nottingham, England, in 1392 in Pipe Rolls, and the Freedom Rolls list an Anthony Ibberson at York in 1596. An Ibberson girl was lady in waiting to Mary, Queen of Scots.

While Joe was waiting in a line of tourists in Florida during the 1970s, he struck up a conversation with a couple from Huntingdonshire County. When Joe asked if they knew any Ibbersons, they replied that many Ibbersons were still living in Huntingdonshire County, and their buildings remained in good repair. Amazingly, they were mostly engaged in forestry and raising livestock. Could a predisposition toward forestry have been embedded in the Ibberson genes?

CHAPTER 3

Growing Up in Lykens

The small town of Lykens is about twenty-three miles north of Harrisburg, the capital of Pennsylvania. To the west of Lykens lies a large, fertile agricultural valley. Several forested ridges are to the south, interspersed with clear streams in narrow valleys. Short Mountain is north of the town, at the westernmost end of the anthracite coal region. Coal had been discovered on Short Mountain in 1825, seven years before the first house was built in Lykens by Joel B. Ferree. A railroad, believed to be the fourth in the United States, was built in 1834 to carry coal westward to the canal at Millersburg.

Lykens Hotel, which Joe's father helped build about 1925. Joe remembered it as being at least twelve stories tall.

Miners' Deposit Bank, where Joe lost the first $100 he had saved by delivering papers and picking blueberries.

The first recorded pioneer in the valley, in 1755, was Andrew Lycans, for whom Lykens was named. Lycans and two companions were attacked by Indians a year later, and Lycans died of his wounds. Evidence has been found of an Indian village on the site of Lykens. An Indian trail extended along a creek into an area known as "the Glen," above which are high rocks that served as lookouts.

Many of the early settlers were Pennsylvania Dutch farmers. Various small ethnic groups came later—Poles, Ukrainians, Italians, Irish, Germans, and Welsh—attracted there by coal-mining jobs and the economic growth that followed. Joe learned how to get along with all sorts of people and became familiar with their accents, culture, and customs. This experience stood him in good stead later in life.

Lykens was a fabulous place for a boy to grow up. The residents were friendly and thoughtful, and the town had some large, well-kept swimming pools and athletic fields. The Glen Park, with a swimming pool built by the community about 1919, was a favorite recreation spot for residents as well as vacationers from afar, offering swimming, camping, boating, picnic pavilions, and a small arcade with a dance hall. There were band concerts on Sunday afternoons and at special events. West Boro Park also had a swimming pool and a dance hall. Most of the kids in Lykens practically lived in the parks during the summer, swimming during the day and watching movies or just hanging out in the evenings.

Joe spent several weeks every summer on his uncles' farms. He helped feed the livestock, milk the cows by hand, load and haul wheat sheaves to the barn for thrashing, shuck corn, and pick peaches from a thriving orchard. He enjoyed helping and learned a lot too. They all turned in to bed early after a hearty dinner and got up at four o'clock the next morning, ready for another day.

Joseph did not play a musical instrument, saying, "A turkey caller was as close as I ever got to a musical instrument, and I was not very good at that." However, he developed an interest in music that remained with him for life. A fellow student of Joe's at Lykens Public School went on to renown in the music world. After Les Brown became famous, his band performed many free concerts in Lykens.

The Glen swimming pool was a great local and tourist attraction.

Not long ago, when the town of Lykens was considering selecting its most outstanding native, Brown and Ibberson were most frequently mentioned.

When he was just five years old, Joe started to take hikes in the nearby forests with his father, older brothers Henry and Ray, and the Ferree brothers Miles and Pete, who taught him to fish, hunt, and trap at an early age. Joe learned from these experts how to tie flies and fly-fish. Because there were few fly fishermen, they had a big advantage and were often successful while those using bait endured disap-

The Lykens Park with this swimming pool also was very popular.

Above: *The Lykens School, where Joe attended all twelve grades from 1922 to 1934, had excellent teachers.* Left: *Joe estimates he passed through this entrance of the Lykens Public School about 9,600 times.*

pointment. These frequent excursions strengthened Joe's emotional attachment to the forests, which he came to love and respect.

As Joe's fascination with forestry grew, he took some long camping trips with Miles, Charles Furlong, and Gilbert Bowman, friends who were somewhat older and astute woodsmen. Joe gained a tremendous amount of outdoor knowledge from them. They would pack their homemade canvas knapsacks with food and hike over three ridges and two valleys to Clark's Valley. They carried lightweight and inexpensive food such as lima beans, oatmeal, and raisins and supplemented them with fish and game. In this wild, remote valley, the four young men enjoyed excellent fishing for brook trout and pickerel. There also was an abundance of wildlife, including beavers, mink, and bobcats.

Miles was an intelligent, analytical observer of nature and an excellent woodsman. Some years after graduating from high school, he had managed to save enough money to attend forestry school at Penn State. He graduated with a degree in forestry in 1942 and went on to join the faculty of the New York State College of Forestry, specializing in forest economics.

Gilbert Bowman, another fine woodsman, applied to be in the first class of a Game Commission school that trained top-level law enforcement officers, and was accepted despite intense statewide competition among applicants. He became a land manager of Pennsylvania game lands.

Charles Furlong was good at identifying flora and fauna and continually taught Joe, Miles, and Gilbert about the plants and wildlife around them. Through a football scholarship and with financial help from his mother, he entered Lebanon Valley College, where he became a successful fullback. After graduation, he went on to become a botany teacher. His son studied forestry at the University of Maine, and like his father, he also played fullback on the football team.

Sometimes these friends would track honeybees to their tree and remove the honey for a real treat. Hunting for honeybee trees was a fascinating sport, done with a simple technique. They first attracted bees with sugar water and a few drops of anise on a piece of paper. The scent of anise lured the bees, which then filled up on the sugar water. The bees would fly away in higher and higher circles, and then finally made a beeline in the direction of their hive in the beetree, until they flew out of sight. Pursuit of the bees required several repetitions of the procedure, moving the bait repeatedly along the flight path until the bee tree was found. The boys sometimes dusted the bees with flour before releasing them, so that when they returned, the elapsed time for the round-trip could help indicate the distance to the hive.

One such honey tree was three miles away, and it took them two days to locate it. But the reward was sweet. The boys found the beehive in a large chestnut oak and at first thought it was too large to fell with an ax, the only tool they had with them. Joe swung the ax hard at the tree in disappointment. But as it struck the trunk, amazingly the ax penetrated into the decayed interior. Within a few minutes, they were able to fell the tree and quickly extract big gobs of blueberry honey before the stunned bees could sting them.

Joe also trapped all kinds of animals with his cousin and classmate Robert Ibberson, the son of Uncle George. The two boys trapped raccoons, muskrats, beavers, foxes, weasels, opossums, and occasionally an unwanted skunk. Joe loved these outings for hunting, fishing, and camping, and he particularly loved the forest. "They were fabulous days," he says.

Brother Henry claimed that as a boy, Joe had a tree growing out of his head. The imaginary tree materialized during interviews in 2004.

CHAPTER 4

Earning Funds for College

Through discussions with his older friends, Joe came to realize that he would need considerable funds to attend forestry school. There were no government loan programs or other types of financial assistance in those days, and scholarships were scarce and small. It was time to get to work and start earning some money.

At the age of thirteen, Joe took over a newspaper route delivering the *Harrisburg Telegraph* daily to ninety-six customers, which enabled him to become acquainted with just about everybody in Lykens. Each customer received six deliveries a week and was charged 10 cents. Of this, Joe paid the newspaper 6 cents per customer, leaving 4 cents for him. Not everyone paid on time, however, so his anticipated weekly profit of $3.84 sometimes fell short. Whenever the coal miners were on strike, the money he collected would hardly pay for Joe's monthly bill. But for Christmas, Joe's customers gave him generous gifts.

Still, funds for college were not building up fast enough, so Joe took a second job at George Eby's wholesale confectionery and tobacco store, where he spent evenings and Saturdays stocking shelves. He frequently went along on delivery trips to stores in the Lykens Valley and helped carry orders into the stores. For this job, Joe was paid $2.50 per week. To impress Joe with this beneficent wage, proprietor Eby would jingle five half-dollars from hand to hand about ten times and then hand them over. Joe was irritated by this maneuver but politely kept his peace.

Even with his two jobs, Joe was not earning money for college fast enough to suit him, so he started to pick blueberries during the summer every day except Sundays. Two Ukrainian boys, George and John Sass, rattled their berry buckets under Joe's bedroom window at four o'clock each morning to awaken him. After a quick breakfast, he joined his two friends for a two-and-a-half mile walk in the dark along an old railroad bed. Their destination was Sheep's Head, a hill on the north side of Berry Mountain. By the time they arrived, it was light enough to pick the berries. Joe picked into a quart container and emptied it when full into a twelve-quart pail. They ate some berries, as boys will do, turning their hands and tongues blue. Usually he had picked twelve quarts by noon and carried the brim-full pail on the long journey home. They picked every day while the berries were ripe. For a twelve-quart bucket, he was paid $1.50 by the local baker, Bill Samson, who could sell blueberry pies as fast as he could bake them.

Joe deposited his earnings from the three jobs in the Lykens Miners' Deposit Bank. As the blueberry season was closing and school was about to start, Joe's bank account held $97. To bring the total to $100, Samson agreed to purchase huckleberries, which Joe picked for two days after the blueberries were done. George Sass accompanied him, as Joe was not permitted to go alone.

What a milestone! By the end of the summer, Joe had earned $100 toward a college education. Several times each day, when Joe passed the Lykens Miners' Deposit Bank, he would think proudly about his $100 safely inside. But disaster struck a few months later. His savings from an entire summer's work were wiped out totally when the government declared a bank holiday. Many banks were closed entirely and liquidated, and others operated under a federal examiner.

But eventually the banks became regulated and depositor insurance was issued. Joe remained as determined as ever to save for forestry school and started a new savings account at the same bank; the Lykens National Bank, the only other one in town, had been liquidated.

From twelve years of age until he reached seventeen and graduated from high school, Joe continued to work at these three jobs and slowly add to his savings account. He received a good basic education at Lykens High School and realized that this, too, was important in preparing for college.

As a lad, Joe picked coal from a waste pile to help out with home heating and cooking.

Joe worked as a breaker boy under conditions such as these.

After graduating from Lykens High School in 1934, Joe was fortunate to get a job as a breaker boy. Breaker boys stood alongside an endless conveyer belt and picked out the rocks and slate from the coal. The mine at Short Mountain produced red ash anthracite coal, which could be sold at premium prices, especially in New England. The word around Lykens was that they had the best red ash coal. Red tickets were mixed with the coal to identify it for more effective marketing. Red ash coal was subjected to greater heat and pressure during its formative period and therefore has a higher Btu content than other types of coal. While burning, it can be dampened down and restarted several times. It leaves no clinkers, only a red dust with small granules. Red ash coal occurs in several veins in Short Mountain and only a few other places. The work at the coal breaker was boring, but the pay was $4.62 a day, which was the union rate for work outside of the mines.

Joe had to join the United Mine Workers Union, Local 1062, and pay the required dues. At a local union meeting to elect national officers, Joe noticed that the only name on the ballot for president was John L. Lewis. Over the years, Joe had heard and read many accounts about the dictatorial Lewis practically closing down America with long strikes at coal mines. These strikes had ripple effects on the steel industry, transportation, and energy industries. Furthermore, he had already lived through the hardships of long strikes called by Lewis. In those days, the strikers got no money from the government or the union and had to survive by credit extended by local storeowners. Joe walked up to the local president, who happened to be his uncle Slim Boden, and asked if he could vote for someone other than Lewis. His uncle advised him in a hushed voice that nobody voted for anyone but Lewis. Joe wouldn't vote for Lewis even though everyone else did, so he didn't vote at all.

The work at the coal breaker lasted only for six months; the profitable supply of coal suitable for the breaker had been exhausted. Employment opportunities in the Lykens area were scarce, as it was the midst of the Great Depression. In fact, most families had no one gainfully employed. So Joe and his friends, being unemployed, took a long camping trip to his beloved Clark's Valley. But he was unable to relax as he usually did, as getting sufficient funds to attend forestry school was foremost in his mind.

Finally, Joe's uncle, George Ibberson, suggested they start an independent coal mine, as the coal company that owned the land had no objections. So Joe became an independent miner, also known as a "bootlegger." Seventy years later, Joe still has coal flecks from mining embedded in his skin. The specks stay there indefinitely unless removed. The hands of some coal miners turned blue-gray forever from coal particles.

Uncle George chose a location on the south side of Short Mountain, opposite Lykens, where he saw evidence of a little vein of coal coming to the surface. The vein was five feet thick, with solid rock above and below. The men started a tunnel in the vein, which had a thirty-degree downward pitch. This required working all day in a stooped-over position. Carbide lamps attached to miners' hats provided light. They cut timber nearby and used it to support the roof of the main tunnel. Most of the coal already had been removed from their bootleg mine by the deep miners digging up from below. Any remaining coal worth mining would come from side tunnels dug at right angles to the main tunnel.

Joe and his uncle drilled holes by hand into the coal, filled them with dynamite, and set off the charges to loosen the coal. Then they had a long wait until the fumes and dust cleared. They shoveled the coal into rubber-tired buggies using hand scoop shovels, and then pushed and pulled them by hand to the main tunnel. At the mine entrance, an old automobile motor with an attached winch and cable pulled the loaded coal buggies to the surface. There the coal was dumped into a chute and slid down into a bin along a truck road. When a full load was ready, they would call a trucker to come for it. George's son Robert did most of the outside work.

Right: *Joe worked as a miner with his Uncle George, inside a coal mine such as this.* Below: *The Short Mountain Breaker, where Joe worked as a breaker boy.*

Joe's mining partner, Uncle George Ibberson, also was chief of the Lykens Fire Department and President of Lykens Borough Council.

One day Uncle George and Joe were drilling a hole for blasting at the face of one of the side tunnels. Suddenly they heard unusual rumbling, crackling, and creaking sounds, and some small stones started falling. Alarmed, Joe immediately dropped his drill and started running full speed toward the main tunnel and daylight to escape a possible cave-in. Then the thought struck him that Uncle George might be trapped and he should attempt to rescue him. As Joe slowed down a little and looked over his shoulder, there was his uncle right behind him. Out of breath, they broke into daylight at the mine entrance and fell to the ground, exhausted. For quite a while they waited and discussed the situation, then decided to reenter and assess the damage. No cave-in had occurred, just a new crack in the otherwise solid rock roof. So they placed several heavy timbers on each side of the crack and returned to work.

Joe enjoyed his jovial uncle and was fortunate to be working with him, as he had been superintendent of all the rock tunnel work for the deep mines near Lykens and Big Lick and knew a lot about mining. He understood the complexities of timbering for safety inside the mine and had great strength for setting large timbers. He was an expert in drilling and blasting to loosen the coal and thoroughly understood electric circuits that set off the dynamite. But unfortunately, they were not making much money because of the low price of coal and their low production, as the deep miners had already removed all the coal that was easy to mine.

Somehow Joe had to find another way to earn money for college. Sister Ruth and her husband, Harry Stutzman, suggested a possibility. They were close social friends of the operators of the Lykens Shirt Factory and arranged for Joe's employment in the collar and cuff department. Collars and cuffs, which had three layers, were starched by one machine and pressed by another. Joe operated a starching machine and was on piecework, getting paid a fixed amount per shirt. He worked fast to boost his income, working an eight-hour shift and also in the evenings. Some weeks he made over $30—in those days considered a very good income. Soon Joe was promoted to foreman of the collar and cuff department, thus being responsible for keeping all the materials in order and the work organized.

The factory was unionized, so Joe had to join the Amalgamated Clothing Workers of America. His reaction was, "Talk about a far-left outfit, all were communists from A to Z." Highly placed officials from New York City gave speeches at the union meetings about marching on Washington to take over the government. They accused the manufacturers of wringing every last drop of sweat from the workers and demanded higher pay to reward them fairly. The factory owners responded that this would cause them to lose money; they even turned over their books to the union as proof. But the union replied that they did not believe the books and called a strike.

Thereupon the owners threatened to move to North Carolina so that they could stay in business. To prevent the move, the union picketed and camped beside the factory around the clock. After about six months, when the picketing had gotten lax, a train pulled onto a siding alongside the factory, and all the machines, which had been unbolted from the floor, crated, and stored near the loading doors beforehand, were loaded and transported to a new location near Morehead City, North Carolina.

The owners offered Joe a job at the new factory, but with lower pay. Joe decided not to move. Pennsylvania was his state and always would be. He had a forestry mission to accomplish, one that had been inspired by Dr. Joseph Rothrock.

After a time, the union did persuade a manufacturer to reopen the shirt factory. The new owner had to replace all the machines that had been shipped to North Carolina. Lewis Hoffman, a Lykens businessman, arranged for Joe to get back his previous job. But the terms were a shock. The pay was only 25 cents per hour, and the union would not permit any piecerate work. The weekly income would be $10, from which were subtracted social security deductions and union dues. What a reduction from the more than $30 a week Joe had been making! The union had accomplished exactly the reverse of its intention. The new owners appreciated Joe's efforts and acceded to his request to work overtime in the evenings, though the pay would still be 25 cents per hour.

CHAPTER 5
Ranger School

Though he had not yet saved up enough money, Joe was eager to begin his forestry education. He submitted an application to enter Penn State's two-year forest ranger curriculum in 1937, without telling his parents or asking for their permission. His first year would be at Mont Alto, spent together with the freshmen of the four-year baccalaureate program, and he would remain there for summer courses. For his second year, he would be at State College, taking mostly senior-level courses. Most of the credits earned in the two-year program could be applied later toward a four-year bachelor of science degree in forestry. Joe's plan was to get employment after completing the ranger program and save enough money so that he could complete the remaining two years required for the bachelor's degree.

When his application was accepted, Joe had to face the problem that he did not have enough funds for even the first year. He worked all the evening hours possible at the shirt factory, but the 25 cents hourly pay rate did not add up nearly fast enough. The owner of the shirt factory tried unsuccessfully to talk Joe into not going to college and just keep working. Joe would hear none of it, as he was determined to pursue his dream.

About two months before the date of enrollment at Mont Alto, Joe informed his parents that he had been accepted but did not have even half of the funds needed for two years. They agreed to help to the extent that they could. Their desire and intention for their eldest son, Henry, to become a forester no longer seemed possible. Henry had gotten married and had other obligations as well. Joe and his parents together placed $650 in a checking account for him, which was their estimate of all the costs for the first year, including a few dollars of spending money. At the end of the year, $30 remained after all expenses had been paid. Joe's many years of frugality no doubt had helped him develop wise spending habits.

At long last, his dream was beginning to come true. Joe reveled in the academic year and summer at Mont Alto. It meant so much to him that he still has his college matriculation card to this day. The second day at Mont Alto, however, he had a flash of homesickness. It was quite an unexpected and unsettling feeling. After all, he had been away from home often on camping trips to Clark's Valley, and he and his father had made many one-week fishing trips to Penns Creek near the Poe Paddy Mountain tunnel. But he had to buckle down, for it was three years since he had graduated from high school and he was somewhat rusty at studies. It took him a month or more to shift into high gear, but it helped that his ranger classmates and the four-year students were friendly, studious, and helpful.

The professors at Mont Alto were outstanding, and they fostered conditions conducive to learning a great deal about forestry. He was especially impressed with Professor M. K. Goddard, who gave a solid hour of instruction at every session. Beside the courses the rangers took with the four-year freshmen, they also had some additional practical courses. One of these was surveying, which turned out to be fortuitous when Joe applied for his first job.

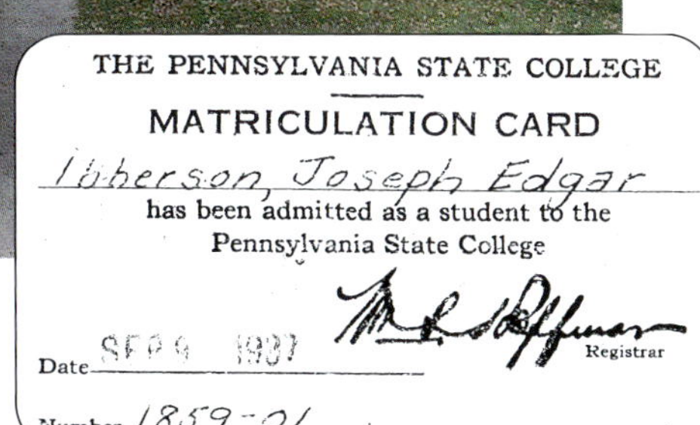

THE PENNSYLVANIA STATE COLLEGE

MATRICULATION CARD

Ibberson, Joseph Edgar

has been admitted as a student to the Pennsylvania State College

Date SEP 9 1937

Registrar

Number 1859-01

Above: *Wiestling Hall at Mont Alto, where meals were served.* Right: *Matriculation card of Joseph Edgar Ibberson, September 9, 1937.*

H. Norton Cope, who was director of the Pennsylvania State Forest School at Mont Alto, also taught dendrology and silviculture. He was an excellent teacher who was amusing and good at practical forestry. Mont Alto is a unique area for the study of dendrology. It is near the northern limit of the natural range of several southern tree species and near the southern limit of many northern species. Because of this overlap of species distributions, students can learn about a large number of tree species. In the summer session, the students already were making forest inventories and management plans. These were of particular interest to Joe, and he planned to specialize in forest management if he ever got the opportunity.

Joe earned $8 a month while at Mont Alto, doing janitorial work for Rook Smith, the building superintendent. He spent most of these earnings on weekend visits to Waynesboro and Chambersburg, where he went out for dinner and shopped for necessities.

Joe played baseball and two-hand touch football. During one game, his quarterback called a fly pattern in which Joe would run as fast as possible to receive a pass. He took off at full speed, with defender William J. Musso from Lynn, Massachusetts, matching him stride for stride. After a long run, Musso bumped Joe hard, changing his direction to one side. Joe never saw the football until it miraculously came over his shoulder and into his hands for a touchdown, the only one of the game. Musso was furious, shouting that Joe was just plain lucky. Joe shot back confidently, "Oh, I make such catches all the time."

Of all the friends Joe made at Mont Alto, the only two he was still in touch with as of 2004 were John C. "Jack" Good and Gilbert G. Gault, both four-year students who graduated in 1941. Jack was severely injured during World War II but recovered and became vice president of the Bartlett Tree Expert Company. Gilbert became a gunsmith. Roland Mayer from Allentown was Joe's roommate at Mont Alto; he later worked for AT&T.

In the fall of 1938, Joe moved to State College and registered at Penn State for his second year. He lived at a boardinghouse for $3 a week at 316 South Allen Street, operated by Gordon and Eva Van Tries. Across the street, at the Harder home, he had good meals at the reasonable weekly rate of $5.

It was a real shocker for Joe to discover so many other important professions represented at the university. It was a broadening experience. At Mont Alto, the studies were oriented to forestry, and Joe had been preoccupied with the importance of forests and foresters. Now he was exposed to students and professors from many other professions—sociology, chemistry, physics, art, music, horticulture, beef and dairy farming, wildlife management, ichthyology, economics, and journalism—all of which had a role in making the world a better place. Joe realized this broader base of knowledge was necessary for success in the real world.

The Green Shack, also known as Fergie's Woodshed. Ibberson attended the last forestry classes held there in 1939.

Joe's year at State College was enjoyable and provided him the opportunity to learn a tremendous amount about forestry and related subjects. He became acquainted with hundreds of people from all parts of Pennsylvania as well as other places. Joe believes he got to know more people at Penn State than all those who lived in his hometown of Lykens, and this enlightening experience in getting along with all kinds of people was helpful when he started his career.

While attending college, Joe found employment with Dr. Wallace E. White, professor of wood utilization, keeping his wood specimens in order for identification tests. About fifty two-inch-square specimens, each with a letter and number that identified the tree species, were mounted on pieces of plywood. Joe had to keep all these wood samples in order so that the tests could be graded properly, and for this he earned $32 per month. Joe took this wood technology course along with the other students, and in the final exam he identified 100 percent of the specimens correctly. Dr. White opined that working for him really paid off, to which Joe replied confidently that he would have made 100 even if he had not worked for him.

Joe's roommate at State College, Robert F. Hilton, had an automobile and went home to Chester County nearly every weekend. Joe helped him with his homework, and for this Hilton let Joe use his car occasionally. Joe and his friend Joseph A. Peel, an education major also from the coal region, borrowed

Ibberson upon graduating from the Ranger Course in 1939.

Hilton's car in order to date nurses who worked at the Lewistown Hospital. Peel was "big man on campus," president of the student union and a starter on the football team.

The 1939 forest ranger class had its forestry courses in Fergie's Woodshed, also known as the Green Shack. This original forestry building was named for the former head of the Forestry Department, John A. Ferguson. The two-story building was constructed entirely of wood, including the frame, roof, walls, floors, and siding. Joe considered it a real tinderbox and always had several emergency escape routes planned. A new forestry building was under construction at the time, and Joe helped Dr. White move into the new building when it was completed. At the time, it seemed so immense that Joe never dreamed some day it would be too small and another new building would be needed. As it turned out, Joe would live to see four forestry buildings on campus in his lifetime.

CHAPTER 6

First Professional Employment

When he graduated in 1939, Joe was the only student from his ranger class or the four-year class with an offer of employment. The job materialized through an interview arranged by Joe's sister Ruth. She and her husband, Harry Stutzman, were regular social friends with Pennsylvania State Treasurer Ramsey Black, who later became Assistant Postmaster General under President Franklin D. Roosevelt.

Black interviewed Joe at his office in Harrisburg. After the interview, Black made a telephone call and sent Joe several blocks away to the offices of a very large engineering firm, Gannett, Eastman, Fleming, Cordurey, and Carpenter. Donald Carpenter interviewed Joe and asked questions, particularly about the surveying courses he had taken. Carpenter wrote down Joe's State College address and promised to contact him if the firm had any openings for someone with his education and experience. He also inquired whether Joe knew Walter and Katherine Zerby, a wealthy socialite family in Lykens. Joe replied that he had sold blueberries to

Joe's sister Ruth helped him secure his first professional job through her friends.

Katherine at one time and had been invited to several parties at their home while attending Penn State. Joe later learned that the Carpenters and Zerbys were close friends.

A short time later, Joe got an unexpected but most welcome surprise—a job offer. He was instructed to report in two weeks, May 10, 1939, to a construction site in Clark's Valley in Dauphin County, Pennsylvania—the same Clark's Valley where Joe and his friends had spent so many happy hours. But he was to report for work three weeks before graduation.

Corewall of the dam under construction. Grout forced under pressure through its many pipes sealed any cracks underneath to prevent seepage.

The dam overflow and retaining walls being built.

Joe thought that in his reply to the engineering firm, he should request a delay in his starting date. But when he showed the letter to Victor Beede, head of the Forestry Department, and several faculty members, they allowed him to leave school three weeks early to start his new job.

In Clark's Valley, a very large dam was to be constructed, creating a reservoir five miles long connected by a twenty-mile pipeline to Harrisburg. The impoundment was for the Harrisburg water supply. Joe reported to the engineer's office at the appointed time. The resident engineer, Hank Porter, appeared to be disap-

Concrete floor of the spillway being constructed.

The intake tower can withdraw water from the level that has the best quality.

Aerial view of the six-mile-long Clark's Valley Reservoir.

pointed that this new employee was not an engineer, but he changed his mind before long.

Joe's first assignment was on a four-man crew, surveying some 6,000 acres of the watershed. The other three members of the crew were highly qualified, experienced, and proven engineers. Engineering jobs were very scarce at the time, just as in forestry. After they had worked together for several weeks, one of the engineers asked Joe what he was being paid. He replied $75 a month, which elicited an amazed reaction by the engineers. Joe assumed his pay was lower than theirs and assured them $75 was the actual amount. But he quickly learned that they were chagrined that a trainee who was not even an engineer was making $10 more than their $65 monthly income. Within a very short time, the salaries of the three engineers were raised to $75 a month. Joe never could decide whether he was a hero or villain to them. Joe's $75 salary presumably came about through Don Carpenter's friendship with the Zerbys, who had been involved in Joe's hiring.

Don Carpenter was in charge of the Clark's Valley project for the engineering firm and was an outstanding engineer. On his frequent trips to the field office in Harrisburg, Joe was impressed to hear Carpenter dictate complicated letters about engineering without notes or references, in rapid fire.

Carpenter often arranged for parties to be held near a cottage where Joe was living. It was a beautiful and convenient place for social affairs. Attending the parties were the elite citizens of Harrisburg, including the Zerby family, and some politicians.

Carpenter invited Joe to these parties and to many social luncheons and dinners in Harrisburg. Joe thought he may have been treated like a son by Carpenter, who had two daughters but no son. Some of Joe's friends, his sister Ruth, and Katherine Zerby suggested that Joe should pattern himself after Carpenter, regarding him as his role model. Jokingly, Joe replied that he was already like Don Carpenter and had been even before he met him.

As work on the Clark's Valley survey continued, Joe was told to spend rainy days studying specifications for all parts of the project, including the dam, spillway, water intake structure, control house, water treatment plant, two stone dwellings, roads, and other, lesser components. The specifications and accompanying blueprints explained to the last detail how the work was to be accomplished, what materials were to be used, and all related data needed by the engineers and contractors. Joe studied and became familiar with more than 100 volumes of specifications and hundreds of blueprints. Instead of objecting to the extra work, Joe realized that he was getting paid for a course in civil engineering.

Ibberson eventually was promoted to inspector. Now he spent most of his time observing the work of the contractors to make sure it was in accordance with the specifications and blueprints. At the end of each day, he wrote a report and submitted it to Hank Porter, the resident engineer, who coached Joe about what to look for and which contractors might create problems. This work as an inspector carried a lot of authority and respect.

The complexity of the engineering involved in building the dam, intake tower, and control house astounded Joe. It was a tremendous opportunity to learn about the engineering features, ways of exerting authority, and working with a wide vari-

Ibberson's forestry crew that worked on the Clark's Valley watershed.

Many visitors on weekends marveled at Clark's Valley Dam, listened to Ibberson lecture, and patronized refreshment stands.

ety of people at all levels. These included the top professional engineers, hydrologists, contractors each with his own agenda, hundreds of construction workers of all kinds, and politicians ranging from respected to disreputable. Joe really liked his work as inspector, and his supervisors commended him for insisting that the contractors adhere strictly to the specifications and blueprints.

On his frequent trips to Clark's Valley, Don Carpenter had long, helpful conversations with Joe. Often accompanying Carpenter was Bill DeHart, a Harrisburg city councilman who had envisioned this reservoir project and made it all possible. Carpenter introduced DeHart to Joe, and they immediately became friends. DeHart occasionally invited Joe to his family home in Harrisburg for dinner.

After Joe had worked about a year on the project, one of the major contractors, Charles S. Moses, offered him a job as assistant superintendent. His pay would be doubled to $150 a month. Joe was surprised by the offer and wondered whether Moses really thought he would be a good assistant superintendent or just wanted him out of the way as inspector. Joe really enjoyed working with the engineers, and his complete loyalty was to Carpenter, whom he told about the offer. After considerable thought, Carpenter replied, "Joe, you go make that easy money, and you can return to the engineers anytime you choose."

Joe accepted the job and found it to be another rewarding experience. Moses placed him in charge of the field office, where he was responsible for the payroll, ordering and receiving materials, and supervising the equipment repair shop.

Charles Moses was an astute businessman, and Joe learned a lot from him. Ordering supplies was a major responsibility. He ordered as many as three railroad cars of cement at a time. The payroll work was complicated yet interesting. Workers included foremen, cement finishers, steel reinforcement fabricators, truck drivers, and laborers. He learned to deal with some rough people.

The equipment repair shop was most challenging, as it was often difficult to get the mechanics to agree on things. The equipment included a fleet of concrete-mixing trucks, special trucks that hauled cement, large dump trucks to haul sand and gravel, smaller trucks used for various activities on the site, a backhoe, a bulldozer, a crane, and several generators. A course in gasoline engines that Joe had taken at Penn State served him well, and his knowledge of equipment earned him respect from the mechanics.

When time permitted, Joe also helped supervise construction activities, where his familiarity with the specifications was put to good use. Many of the employees were from Lykens; in fact, Joe helped some of them get jobs. His brother Henry was employed as a carpenter's helper. This experience later qualified Henry as a carpenter at Indiantown Gap when World War II started and ultimately led to his being placed in charge of all construction at the Carlisle Navy Depot.

As Clark's Valley Dam was being completed in 1940, Joe received another surprise when DeHart and Carpenter offered Joe the position of city forester, with the responsibility for overseeing the entire finished project. The city of Harrisburg was planning to acquire all of the watershed lands, including several cottages adjacent to Joe's residence. He would be in charge of the whole 6,000-acre watershed, the reservoir, the intake tower, the treatment plant, and a testing laboratory for turbidity, chlorine content, alkalinity, and pH—quite a responsibility for a twenty-three-year-old. The two men told Joe he was the logical person, as he had seen everything built and was familiar with all the components, and as a forest ranger he could manage the large watershed. The offer consisted of a $1,888 annual salary and residence rent-free in one of two beautiful stone houses just being completed. Joe took all of one day to think it over before accepting.

At that time, the reservoir was filling up. Soon it would be time to open certain valves to drain the

Ibberson with his limit of pickerels.

impure water from the bottom of the lake. Much of the watertreatment equipment was being installed. Factory representatives were helpful in briefing Joe on how all this new equipment functioned and how to keep it in good repair. Joe hired a crew of fifteen workers to plant trees, do forest improvement work, and perform other duties.

The City of Harrisburg instituted condemnation proceedings against the 6,000-acre watershed lands, mostly owned by the Reading Coal and Iron Company. Being familiar with the boundaries, Joe conducted tours for the Board of Viewers, attorneys for both parties, and other representatives. The Board of Viewers determined the price to be paid by the city, with Joe's assistance in determining the timber values. The final price appeared to please everyone.

Finally the twenty-mile-long pipeline from the reservoir to Harrisburg was sterilized, the treatment plant began operating, and the water quality at upper levels in the reservoir was satisfactory. The treatment plant was in operation twenty-four hours a day and needed at least one employee there at all times. One mishap did occur while Ibberson was chlorinating the water. The HTH powder that Ibberson dispensed left a residue in his hair, which came out in handfuls when he showered. His receding hairline and sparse hair have persisted from that time onward.

The reservoir started supplying water to Harrisburg in 1941. The project was a major accomplishment and was named the DeHart Dam and Reservoir. Thousands of people visited the area on weekends and holidays to gaze at this wonder. On those occasions, Joe would be present on top of the dam near the intake tower to answer questions. So many visitors came that vendors were kept busy supplying refreshments, which they prepared and served from two vehicles.

The treatment plant, dam, and pipeline are still operating in basically the same way as of this writing. Much of the original equipment is still functioning well; only minor maintenance and repairs have been necessary.

During the time when Joe was managing the DeHart Reservoir, he met a politician who would have a pronounced beneficial influence later in his career. Councilman DeHart arranged for several weekend parties to be held at the control house in honor of M. Harvey Taylor, with card games and drinks. Taylor was currently the most powerful politician in Pennsylvania. Any Republican candidate for governor or the U.S. Senate needed his approval to secure party support. He handpicked every Republican-backed candidate in Dauphin County and usually endorsed only highly qualified people. He had such an efficient political machine that just about every one of his candidates was elected. Later he became the president pro tem of the Pennsylvania Senate and held sway there for many years.

Joe soon became a friend of Harvey Taylor, playing cards with him and having many side conversations. Taylor's friendship turned out to be the key to many accomplishments after Joe was employed by the state in 1948.

On a recent visit, Ibberson met Daniel E. Standish, director of the Harrisburg Bureau of Water (at left), and Chad E. Bingaman, superintendent of DeHart Dam and watershed manager.

To Joe, the work at the DeHart Reservoir was challenging and rewarding, yet also entertaining. He was provided with an inboard motorboat to patrol the lake and its shores. Before long, the lake was teeming with pickerel, many about twenty inches long. A fisherman could easily catch the limit within an hour. When bass became more plentiful, the pickerel population declined, but the bass were a lot more fun to catch. Game species were plentiful on the property and attracted hunters. With all these attractions, there were visitors all the time.

The years from 1939 through 1941 were amazingly satisfying to Joe. He already had acquired a lifetime of valuable experiences, or so it seemed to him. And it was hard to account for the good fortune of having so many influential people giving him immeasurable help along the way. Furthermore, a young man with a good job, a car, a handsome house, and a motorboat had no trouble getting dates with the ladies. To Joe, it was a miracle that all this was happening in Clark's Valley, the area he had fallen in love with at the age of thirteen. It was tempting to think at times that life just couldn't get any better.

CHAPTER 7

Military Service

All that changed suddenly. Joe's comfortable life at the reservoir and his aspirations for his career in forestry were disrupted unexpectedly by the entry of the United States into World War II after the Japanese attack on Pearl Harbor in December 1941. A military draft was instituted, and Joe fell into the category of young men who would be called up first. He could have asked for a draft deferment, based on his responsibilities at the water treatment plant, but his sense of patriotism prevailed and he chose to enter the armed services. Winning World War II was important if freedom was to be preserved.

On the morning of August 10, 1942, he boarded a bus at the main square of Lykens with nothing but the clothes on his back and his wallet. He was dropped off at the New Cumberland Army Depot, across the Susquehanna River from Harrisburg, where he was speedily issued Army clothing, personal items, and a duffel bag for carrying his belongings. Decked out in his Army uniform, Joe had to listen to rapid-fire indoctrination lectures that explained everything from military rules and regulations to the dire consequences of consorting with the wrong type of woman. Between lectures, each

Ibberson as a U.S. Army officer during World War II, showing battle stars and other insignia.

draftee was inoculated against a great many diseases. All this continued at a quick pace until eight that night, when he was given an aptitude test with hundreds of questions. It seemed especially difficult after a long, fatiguing day, and he was beginning to have side effects from the injections, such as muscle pains, fever, and confusion. Afterward, he had barely enough energy to stagger to his bunk. He collapsed into bed without changing his clothing and immediately dropped off into a deep sleep.

A half hour later, Joe was roused by a sergeant yelling that the train was leaving shortly. The new buck privates donned their rain parkas and stood in the pouring rain until they boarded the train for a trip to parts unknown. For the next six days, the train traveled west. There were few stops, and no troops were ever allowed to get off. Two coal-fired steam engines pulled the train as it chugged through the six-mile-long Moffett Tunnel through the Rocky Mountains and the passenger cars quickly filled with coal smoke, which soiled the men's new uniforms.

On the sixth day, the Golden Gate Bridge came into sight, and the buck privates realized that their destination was San Francisco. Buses took them across the bridge to an Army post several miles to the south. There Ibberson lived in a barracks with his own bunk and footlocker. The normal twelve weeks of basic training were compressed into eight weeks because of the urgent need for trained soldiers.

The first time Joe was on the rifle range, he established a reputation as a sharpshooter. At his first turn, he hit the bull's-eye in every one of the sixteen shots. The sergeant in charge said, "You must think you're pretty good." Joe replied, "Sure, and I can also do it left-handed." Thereupon he proceeded to shoot left-handed, with the same accuracy. He was duly appointed to be coach of the rifle range. His skill at shooting accurately had been developed while hunting game for many years.

After basic training, Ibberson was assigned to a seacoast artillery unit, to be trained in aiming and firing large naval guns at enemy ships. The Japanese had submarines prowling nearby in the Pacific Ocean and were even invading some islands near Alaska. He also was given exhausting infantry training so that he could fight the enemy in case they overran the coastal gun positions. After two weeks, Ibberson was promoted to acting corporal, as there was a shortage of noncommissioned officers. Now he had a squad to command.

When the eight weeks of basic training were completed, orders came to the men to pack their duffle bags. They were going to the Aleutian Islands. But at the last minute, Joe was assigned to Fort Scott instead, located at the Presidio in San Francisco just south of the Golden Gate Bridge. When Ibberson returned to the barracks, his buddies were happy for him but envious. Fort Scott was regarded as a very desirable assignment. There Joe was housed in an apartment instead of barracks. Fisherman's Wharf was within walking distance, and streetcars gave ready access to San Francisco.

Joe was sent to Fort Scott because of his mathematical abilities, high score on the aptitude test, and surveying experience. Again his forestry education at Penn State had stood him in good stead. Surveying principles were very much involved in the gunnery practices of U.S. Army Seacoast Artillery, which protected against Japanese ships that were lurking offshore.

The gun battery to which Ibberson was assigned had highly advanced technology, with radar installations to locate and track targets, instead of the usual telescopes, and an enormous gear-driven computer that aimed the guns by cables attached to them. If the first projectile missed the target, the splash in the ocean would be detected by the radar. With this information, the computer would adjust the next shot, which usually hit the target.

Because the guns were manned by shifts around the clock, one- to three-day passes could be arranged with considerable flexibility. Ibberson took advantage of this to spend time at the San Francisco financial center, including the stock market. He made some friends there and purchased small amounts of common stock in several companies. Already he was hoping that sometime later he might be able to make major philanthropic gifts. He also took trips to the magnificent forests and parks of California, including the redwoods.

Besides being part of the gun crew, Ibberson continued leading infantry training and maneuvers, as well as conducting daily calisthenics. It was exhilarating to mount the platform, to demonstrate the various exercises with precision, and then to give the commands. The troops would finish with a 400-yard run, and Ibberson often finished first.

After about two months, several of Ibberson's commanding officers suggested that he apply to the Officers Candidate School at Fort Monroe, Virginia. Ibberson liked what he was doing and enjoyed San Francisco and the sights of California. But after a year, he finally took and passed the examination and was soon on his way to Fort Monroe, where the seacoast artillery had been headquartered for many decades.

Ibberson was surprised that he and one other candidate were the only two from the Regular Army to arrive at Fort Monroe. All the others were graduates from the University of Pittsburgh and the University of Alabama who had completed four years in the Reserve Officers Training Corps. Ordinarily that would have qualified them to be commissioned as second lieutenants. But that ruling had changed, and now they had to successfully complete the Officers Candidate School program.

The rigors of the program were as severe as the rumors had predicted. The only other candidate from the Regular Army lasted for only two weeks before being returned to his unit. The workload and assignments were nearly impossible to complete on time. There were daylong and evening sessions from Monday through Friday. At first everyone thought Joe was the G2 intelligence man, because he associated with officers at evening lectures.

Every soldier's uniform, military bearing, and living space in the barracks had to be perfect at all times. Marching and formations had to be performed with West Point precision. The math courses and artillery problems were exceedingly difficult. The program often was referred to as "three months of hell." The real test was determining whether the candidates could withstand the pressure. To survive, the fresh college graduates often asked Ibberson for information and advice, knowing that he was from the Regular Army.

The company commanding officer, Captain Fish, happened to be a forester, and perhaps for this reason, he seemed to take a special interest in Ibberson. He often ordered Joe to carry out various assignments, partly because of his prior Army experience. He designated Ibberson as gun pointer when the 75-millimeter French mortars were fired. He assigned Joe to be in charge of battalion formation and drills. When it was time for the company to return to base at the end of a day of maneuvers, it was usually Ibberson's duty to march them home.

Several of the instructors asked Ibberson if he knew of Dr. Edward A. Ziegler, who had been a director of the Mont Alto Forest Academy before serving in the Army coast artillery during World War I. Ziegler was a legend at Fort Monroe. He had been the leading instructor in mathematics and in solving artillery problems and had a reputation as a superb teacher.

Captain Fish would try repeatedly, though in a friendly way, to find a task that Ibberson was not familiar with and could not do. One day he ordered Ibberson to mount a platform and direct the calisthenics of a rather large group. As he passed Captain Fish at the platform steps, Ibberson gave his best salute and noted Fish's facial expression. Clearly, Captain Fish was thinking, "I got you this time," for he expected Ibberson to be confused when demonstrating the calisthenic exercises. Instructors were required to give commands for movements toward the right while demonstrating them to the left, and vice versa. Fish did not realize that Ibberson had done this on a daily basis at the Presidio. After he had performed flawlessly for about ten minutes, Fish ordered him from the platform and asked Ibberson where he had learned all this. The answer was "From First Lieutenant Vaughn at the Presidio," whom the captain happened to know and respected.

Joe had arrived at the academy at Fort Monroe as a corporal, and he graduated highest in his class as a second lieutenant in the seacoast artillery. Ibberson and another graduate were assigned to a secret project near Virginia Beach, Virginia, where they were told that advanced technology was being developed and tested that would revolutionize artillery gunnery. But after several introductory sessions at Fort Monroe, they learned that the Virginia Beach project was being discontinued in favor of an even more advanced project in California. The reason was that seacoast artillery, a defensive weapon, was becoming of secondary importance in the offensive campaigns in the Pacific and European theaters.

Ibberson was stationed next at Newport News, Virginia, in 1944 and began his training to become a transport commander. There was a shortage of officers who could serve in this position on ships carrying Army troops. A transport commander was responsible for all troops aboard his assigned ship and was completely in charge of all their needs and activities.

Knowing that he would be sent overseas, Ibberson sold the Chevrolet that he had bought new in 1941 for $650. The dealer had claimed that was $5 below his cost. With 35,000 miles on the car, Joe was able to sell it for $3,500. Automobiles had become scarce because much of the manufacturing had been diverted to the war effort. The money was a big help later when Joe returned to college.

After some training and two trips across the Atlantic as assistant to an experienced transport commander, Ibberson became a full-fledged transport commander and was promoted to captain. At times the work was quite demanding. Within a few hours, 5,000 or more troops would board the ship, and he had to be ready for their dining, sleeping, policing, recreation, medical care, law enforcement, and anything else that might come up. This required extensive and detailed preparations.

But there were many rewards as well for a transport commander. After the port of destination was reached and the troops disembarked, usually two to three weeks

Ibberson always arranged frequent entertainment for his military passengers, in this case on the hangar deck of the USS Lake Champlain.

The Navy took advantage of Ibberson's gunnery expertise when the ships were not carrying passengers.

passed before the return trip started. During this time, the Army provided transportation for sightseeing, and Ibberson took full advantage of such opportunities. He climbed the Great Pyramid at Giza, Egypt, and spent considerable time in and around Cairo. He walked the streets of Pompeii, where the ashes from the volcanic eruption of Mount Vesuvius in 79 A.D. had been removed. He climbed the Leaning Tower of Pisa and thoroughly explored Florence and Naples. He enjoyed opera and attended many performances while in Italy. He toured much of North Africa, where he saw hundreds of German tanks partially buried in the sand.

Usually the ships traveled in convoys of up to fifty vessels. The convoys were protected by destroyers, but there were so many German submarines that ships frequently were hit by torpedoes. When tankers were hit, the ocean surface would be ablaze. In one incident, Ibberson's ship continued right through such a fire. In the darkness, it was an eerie sight.

On another trip, Ibberson's ship had passed through the Strait of Gibraltar and proceeded westward about six hours into the Atlantic, when suddenly there was a loud explosion. At the time, Ibberson was having dinner in the officers' dining room, and he felt the ship abruptly rise up and then settle down in the water. Believing that the ship had been torpedoed, he headed for his assigned lifeboat according to instructions. As he left the dining room, he noticed a basket of cookies and filled his pockets with them, thinking he would soon need a lot of energy. He raced to the lifeboat and was surprised to see a ship nearby with a large, gaping hole in its side. Crew members who had survived were frantically lowering their lifeboats and fleeing the damaged ship as fast as possible. Ibberson's ship had not been torpedoed after all; it actually was undamaged and continued to the United States without further incident.

The destroyer escorts that accompanied the convoys in which Ibberson traveled often fired depth charges at the submarines. Sometimes all of the destroyers would converge and fire depth charges at a place where a submarine was thought to be lurking.

Ibberson made twenty-two round-trips across the Atlantic Ocean, most of them going into the Mediterranean Sea and the most distant one to Egypt. His trips started and ended at various ports along the Atlantic and Gulf coasts, including New York City, Philadelphia, Baltimore, Norfolk, and New Orleans. On days when he was off-duty Ibberson frequented financial centers and stock exchanges in the port cities. While at Fort Scott, he often played poker in his spare time, playing conservatively, and usually won. He invested most of his poker winnings in small purchases of stocks and thereby learned a lot about investing. The employees of financial institutions were eager to help a uniformed captain and gave freely of their time. Later, when Ibberson began investing much larger amounts, the experience he had gained in this way proved invaluable.

There were about 8 million U.S. troops in Europe when Germany surrendered. Most of them were no longer needed, so there was a huge outcry to have them returned home as soon as possible. Thus the Magic Carpet Movement was born. Every possible means of transportation was mustered to return the GI heroes to America. Passenger vessels, cruise ships, cargo ships, and just about anything that could move troops was commandeered and pressed into service.

The *Lake Champlain,* a 27,000 ton aircraft carrier, was one such vessel. It had just finished its shakedown cruise and was carrying a full deck of aircraft. The United States was also winning the war in the Pacific, so it made sense to use this ship for the Magic Carpet Movement. Its crew removed the airplanes from the hangar deck and installed bunks five high. They also installed enlarged dining quarters and other necessary facilities to accommodate 6,000 passengers. All the transport commanders thought it would be a plum to be assigned to the *Lake Champlain.* Ibberson was the one fortunate enough to be chosen, having served on other naval ships.

Ibberson had an excellent relationship with the executive officer, Commander Koepke, a former All-American halfback at the Annapolis Naval Academy. Once while the carrier was docked at Norfolk, Virginia, between trips, Commander Koepke invited Ibberson to fly with him in a fighter plane to Philadelphia to see the Army-Navy football game.

The *Lake Champlain* was a fast ship and could cross the Atlantic in four days. On one trip, the carrier broke the all-time record, averaging 32.048 nautical miles per hour. As a member of the ship's company on that occasion, Ibberson was presented with a blue-ribbon award that he eventually displayed with other awards in his office.

After three and a half years of active military service, Ibberson was ready to be discharged in early 1946. World War II was over, and he had served his country when it needed him, so now he was eager to return to his life's mission. The next step was to earn a bachelor of science degree in forestry at Penn State.

When Ibberson completed his last overseas trip at New York City, he was assigned to the Brooklyn headquarters. He reported to his commanding officer, a full colonel, and requested to be discharged so that he could attend forestry school.

Ibberson's certificate showing that he was a member of the ship's company aboard the USS Lake Champlain *during a record crossing of the Atlantic.*

But a point system was used to determine eligibility for discharge, based on length of service, age, and other variables, and Ibberson was not considered eligible.

Ibberson was undeterred. He was determined to enroll at Penn State in the spring semester, which would start soon, so he went directly to the office of the commanding general at the Brooklyn headquarters. He told the general that he had been willing and proud to serve his country and had actually enjoyed and benefited from his military experience. But his lifetime ambition was to become a forester and make a contribution to humankind, so he wanted to return to Penn State to complete his education. The Army did not need him, he said, as there seemed to be an adequate number of transport commanders, and discharging him would save the Army the cost of a captain.

The General signed some documents that would permit Ibberson to be discharged from the Army immediately if he passed a physical exam at Fort Dix. The next day, Ibberson passed the physical and was discharged.

He immediately phoned the Penn State admissions to reenroll. But housing was so tight that he had to show proof of having lodgings before being permitted to enroll. Another phone call, to his former landlord Gordon Van Tries, solved the housing problem. Ibberson rushed directly from his base in Brooklyn to Penn State just in time to start the spring semester.

CHAPTER 8
Baccalaureate Studies at Penn State

When Joe resumed his plan to earn a bachelor's degree in forestry, he had income from the GI Bill for his wartime service and a pretty good bank balance. Upon his return to State College in 1946, he found many new buildings that had been constructed since 1939. He did not converse with fellow students about his wartime experiences, wanting to put them behind him and get on with his life.

The returning veterans generally were more mature and serious about their studies than the younger students. Most of their expenses were being paid under the GI Bill, and they were eager to take full advantage of the opportunities that awaited them. Joe took huge course loads upon his return. During his senior year, he passed fifty-one credits consisting of eighteen courses in two semesters, plus a required surveying course that he passed by examination. A normal load at that time would have been about thirty-six credits in two semesters. Joe adjusted easily to the academic environment and had no problem with any of his courses. In fact, he had improved his skills in algebra while in the military service, realizing that algebra had a variety of applications. He had carried an advanced algebra book with an answer book and practiced until he could successfully work any of the problems in the book. This proved to be wise and an immeasurable help.

Joe's facility in math also created some unusual situations for him in a couple of courses. In a physics class, the professor sent Joe to the blackboard to solve a problem. Joe read the problem, wrote an equation, and then expanded it into a complex, compound equation. By visual inspection, he noted that several of the

Above: *The Forestry Building, later renamed the Ferguson Building, where Joe had forestry classes in 1946 and 1947.* Right: *Old Main, Penn State's administration building while Joe completed his B.S. degree.*

terms canceled out each other. This left the answer, which he wrote under the equation. The lady professor, who had a doctorate, said, "I see Joe was not able to work his problem," and proceeded to fill three blackboard sections with her solution. When finished, she asked Joe if he now understood how to work the problem. Joe, sitting in the front row, replied in a quiet voice, "Your answer is the same as mine." The professor was surprised and fired back, "You must be one of those GIs who studied advanced math in Germany." From then on, Joe sat in a back seat and kept his thoughts to himself. In the end, the professor gave Joe a 3 in the course, which was at the top of the grading scale in use at that time.

Another event involving math occurred in a mathematics course taught by a Professor Kinney, an excellent teacher. He wrote the midsemester exam consisting of four questions on the blackboard. Joe started to work as soon as the first question had been written and was finished several minutes after the fourth question appeared. As he turned his paper in, Professor Kinney asked him if he couldn't solve the problems. Joe replied, "Yes, I am finished." Then the professor asked whether he had checked his work, and Joe said, "I don't have to." Some of the students wondered why he was leaving so soon.

Several days later in a chance meeting on campus, Professor Kinney told Joe he had scored 100 percent in the exam. The professor remarked that he had never seen someone complete a test so quickly and accurately in all his years of teaching or as a student. He would have been convinced that Joe had obtained the exam in advance, had he not made up the problems as he was writing them on the blackboard. Joe's ability to answer the questions as fast as they were written was a unique occurrence and not typical. He really struggled with the final exam and was the last to finish, much to the satisfaction of the professor.

Among the forestry students at Penn State with Joe were Darwin P. Palmer, L. D. "Whitey" Schaller, Henry Wessel, and Robert W. Wilson Jr., some of whom took their meals with Joe at the Penn Haven boardinghouse. Darwin's father-in-law, who was wealthy and lived in Delaware, invited Joe, Darwin, and George Harvey to stay at their home and go deep-sea fishing. George was renowned for his knowledge of fishing and taught fly fishing. When Joe undertook graduate studies later at Yale, Darwin was his roommate for a semester.

Students in forestry and some other majors were well aware of Joe's academic prowess, especially his knowledge of mathematics, and often wanted to discuss course assignments or ask for his help through tutoring. The one telephone in Joe's rooming house was constantly ringing every evening, and whoever answered usually would yell for Ibberson. So many calls came for Joe that eventually a sign was placed next to the telephone stating that Ibberson was not available. Yet he continued to do more tutoring than he really had the time for, if he was to get his own work done.

Joe was especially pleased with the help he was able to give a student from McVeytown. This education major, who was in his last semester, had failed two previous attempts to pass a physics course, even with the help of two other tutors. He needed this course to meet graduation requirements and also to qualify as a teacher, which he desperately wanted to be.

Joe devised a clever way for the student to solve physics problems, enabling him to graduate on schedule, and get a teaching job. The student was proud when he learned from Joe that he came from the same hometown as Dr. Joseph Rothrock.

Some lighter moments also occurred while Joe was enrolled at Penn State. During an entomology field trip, the students were walking in a field along a Scotch pine plantation. It was a rather large class, and Ibberson and Floyd W. Lang were bringing up the rear. The professor picked some clumps of brown material from the pine branches and smelled it, then identified it as insect frass. Ibberson noticed the tracks of a manure spreader on the ground and smilingly mentioned to Lang that the professor was smelling manure, not insect frass. Lang, who was a varsity member of the Nittany Lions football team, muscled his

Professor H. Arthur Meyer, who encouraged Joe to use his influence for making state forest management plans.

way to the front of the group and told the professor, "That's horse s—— you're smelling." Perhaps he was trying to make points with the professor, for he had missed many classes because of football practice. But the professor was not amused.

Joe again worked part-time for Professor Wallace White, as he had in the Ranger program, preparing wood specimens for wood identification tests taken by students. He also dug holes for fencepost experiments that tested their durability.

Joe found most of his forestry and other professors to be good teachers and generally helpful. Forest management was Joe's special interest, so he especially admired Professor H. Arthur Meyer, whose specialty was forest management, and they became good friends. Meyer had become an expert in the subject in his native Switzerland. He would often comment that it was unbelievable and inexcusable that the state of Pennsylvania did not have forest management plans for its 2 million acres of forests. He brought up this topic in the classroom, and also at times when Joe helped with construction work on a farm that Meyer and a colleague owned near Boalsburg. Joe thought that Dr. Meyer probably was challenging him to exert his influence, because Meyer knew about Joe's political friends and his employment status.

When Dr. Meyer asked Joe to take him for a visit to the 6,000-acre watershed of the DeHart Reservoir, Joe also set up a meeting with Bill DeHart. On that occasion, Senator Harvey Taylor happened to be in Harrisburg City Hall. DeHart introduced Meyer to Taylor, who warmly greeted Ibberson. Meyer noticed this and commented to Joe about their close friendship.

After what he had witnessed on the trip, Meyer became insistent that Joe should use his influence to become engaged in making forest management plans for the state forests. Meyer knew of the influence and power of Harvey Taylor and had observed his friendship with Joe. He saw this as a great opportunity to promote his interest and Joe's in forest management. From then on, he never passed up an occasion to bring up the subject. And it only strengthened Joe's resolve to pursue a career in forest management, he hoped in Pennsylvania.

Joe also struck up a friendship with Victor A. Beede, professor of forest management and head of the Department of Forestry. At times Beede asked Joe about his various jobs, experiences, and plans for the future. He agreed with Dr. Meyer's

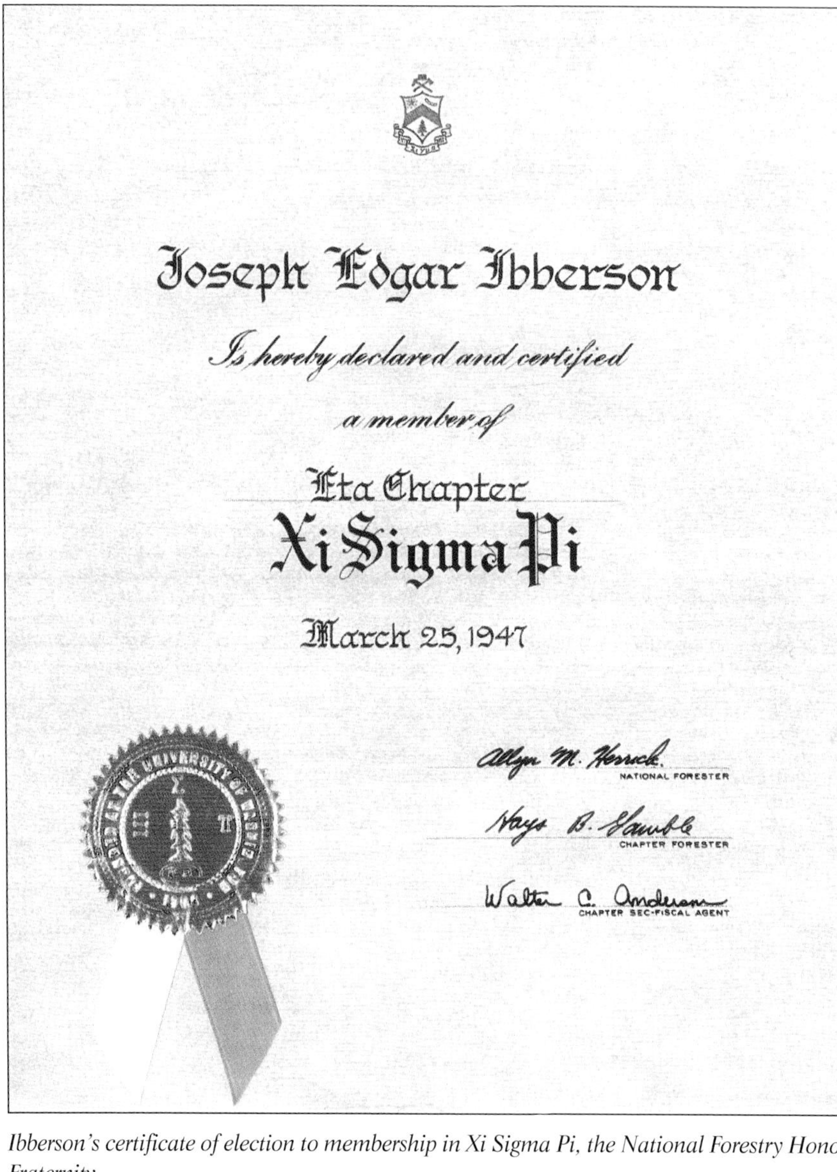

Joseph Edgar Ibberson

Is hereby declared and certified

a member of

Eta Chapter

Xi Sigma Pi

March 25, 1947

Allyn M. Herrick
NATIONAL FORESTER

Hays B. Gamble
CHAPTER FORESTER

Walter C. Anderson
CHAPTER SEC-FISCAL AGENT

Ibberson's certificate of election to membership in Xi Sigma Pi, the National Forestry Honor Fraternity.

recommendations for Joe's career, and suggested that Joe enroll at Yale to study for a master of forestry degree, specializing in forest management. Beede had graduated from Yale in 1912 and was currently president of the Yale School of Forestry Alumni Association. Joe realized that Dr. Beede was recruiting for Yale and appreciated his advice and support. He also persuaded his classmate Darwin Palmer to enroll at Yale.

CHAPTER 9

Graduate Studies at Yale and Finding Employment

Joe applied for admission to the Yale School of Forestry in 1947 and was promptly accepted. The opportunity to become acquainted with top students from other countries made the experience all the more interesting. Joe learned that several classic Yale forestry professors were nearing the end of their careers, and he was fortunate to be able to study with them. Among these were H. H. Chapman, professor of forest management; Ralph Hawley, professor of silviculture; Harold Lutz, professor of forest ecology; and J. S. Boyce, professor of forest pathology. All in all, Yale was a great confidence booster for Joe, after Penn State had prepared him well to build on his foundation in forest management.

The forestry library at Yale was especially complete, well organized, and useful. The librarian, who had a doctorate degree, and had been there many years, kidded Joe in a friendly way that he knew more about the library than she did. Joe used the library frequently and wrote many papers there. One of the papers was for an advanced course in aerial photos. Joe was convinced that aerial photos would be most useful for managing forest resources, and that knowledge proved to be beneficial throughout Joe's career. Another paper was on the important infectious diseases and insect enemies of American elm; that paper also turned out to be helpful.

For one course, Joe wrote a forest management plan for Dauphin County, Pennsylvania, where he had grown up. When a classmate from Rhode Island made fun of him for writing about a single county, Joe countered, "Do you realize there is more forestland in Dauphin County than in your entire state?" Another paper, a very lengthy one, was on community forestry; the term "urban forestry" had not been coined yet. This one turned out to be very constructive too, years later when Pennsylvania became active in urban and community forestry.

George H. W. Bush was in the Yale Class of 1948 and sometimes ate in the same dining commons where Joe took his meals. They spoke with each other on several occasions. One Saturday, they left the dining commons together after lunch. As it was pouring down rain, they debated whether to attend the football game that afternoon. They realized they probably would get soaked at the Yale Bowl. Bush stated forcefully that if the football team had to be there, he felt obligated to attend too; that was typical of his sense of commitment. The heavy rains continued throughout the afternoon, so Joe decided instead to catch up with his studies.

Near the end of his days at Yale, Joe and three forestry student friends decided to go to New York City. The Belmont Stakes were that weekend, so they took the train to Long Island and spent Saturday afternoon at the races. Joe had read about Citation, a racehorse that appeared to be very fast, with a lot of stamina. To the amazement of his friends, Joe bet $500 on Citation to win. They went to the rail to watch the race, in full view of the starting gate. Their eyes were glued on Citation in post number one. As the horses exploded out of the starting gate, Citation stumbled and lost his stride. They all thought Joe would surely lose his $500. But Citation was uninjured, and Eddie Arcaro, an outstanding jockey, soon had him back in stride. Eventually Citation got the lead and won at two-to-five odds, so Joe collected $700, pocketing a $200 profit.

The $500 wager became an item of conversation back at the Yale Forestry School. Over subsequent years, the exaggerated story apparently became a legend. On several occasions, when Joe met students who had attended Yale, he heard, "Oh, you are the Ibberson who paid your way through Yale by playing the horses." Actually, that was the only bet that Joe made while attending Yale, but he thought it did make a good story.

At graduation ceremonies in 1948, when Joe received a master of forestry degree, Yale University bestowed an honorary doctorate of laws on General Dwight D. Eisenhower. At a luncheon for graduate students, Ike gave a short speech at a university dining hall, even though he had accepted the invitation with the proviso that he would not be called upon to speak. The general's remarks were very inspirational. Ike emphasized that to be a successful leader, you must completely forget about yourself and instead think about others. Forestry Dean George A. Garratt arranged for Joe to have a short conversation with the general, as Joe had told Garratt about his family's connection to the Eisenhowers.

During their encounter, Ike commented that his father was an avid reader with insatiable curiosity. He had a logical, inquiring mind and applied precision to

PRAESES ET SOCII
UNIVERSITATIS YALENSIS
IN NOVO PORTU IN RE PUBLICA CONNECTICUTENSI OMNIBUS
AD QUOS HAE LITTERAE PERVENERINT SALUTEM IN DOMINO
SEMPITERNAM NOS PRAESES ET SOCII HUIUS UNIVERSITATIS
Joseph Edgar Ibberson
SECUNDI HONORIS ACADEMICI CANDIDATUM AD GRADUM
TITULUMQUE FORESTARIAE MAGISTRI ADMISIMUS
EIQUE CONCESSIMUS OMNIA IURA PRIVILEGIA INSIGNIA
AD HUNC HONOREM SPECTANTIA
IN CUIUS REI TESTIMONIUM HIS LITTERIS UNIVERSITATIS
SIGILLO IMPRESSIS NOS PRAESES ET SCRIBA ACADEMICUS
SUBSCRIPSIMUS a.d. X KAL. IUL. MDCCCCXXXXVIII

SCRIBA

Charles Seymour
PRAESES

Ibberson is proud of his sheepskin diploma in Latin and feels fortunate to have been accepted for graduate studies at Yale.

everything he did. These same qualities have been attributed to Ibberson. Ike also spoke of his father's favorite Pennsylvania Dutch dishes, such as fried mush, "pudding," and schnitz und nepp (dried apples cooked with ham and dumplings). Ibberson's mother often served these three dishes.

Graduation was a time of great anticipation. Most of the forestry students seeking employment in the months before graduation had job offers. Federal and state agencies all were expanding forestry activities in 1948. Wood-using industries were depending more on professional foresters. Interestingly, two of the students graduating in Joe's class decided to study for the ministry. That may not be as strange as it may seem, for the callings of religion and forestry both are concerned with the future of humanity.

Upon his graduation, Joe had several opportunities for employment. He was appointed city forester of New Haven, Connecticut. He had interviewed for this position but never actually agreed to accept it. His lifelong friend Miles Ferree, who was on the forestry faculty at Syracuse, New York, told him about an open position there and offered to help with his application. Also, as a returning veteran, Joe was still officially an employee of the City of Harrisburg. City officials informed

Joe that they had a big surprise for him when he returned. But he never asked what the surprise was, so to this day he doesn't know what it might have been. Actually, he was holding out for an opportunity to pursue his vision: to originate the first forest management plans for all of the state forestlands in Pennsylvania.

As the time for a decision was closing in on him, Joe received what seemed like a message from heaven. Dean George Garratt of the Yale Forestry School showed him a letter from James N. Morton, a 1916 graduate of the Mont Alto Forest Academy and now chief of the Division of Forest Land Management of the Pennsylvania Bureau of Forestry. The letter stated that he was recruiting a forester to work on forest management plans for the state's 2 million acres of forestland. Garratt thought this should be of interest to Joe, because he was from Pennsylvania and had specialized in forest management.

Joe immediately telephoned Morton and made an appointment for an interview in Harrisburg. After the interview, Morton decided to hire Ibberson if he could get political sponsorship; foresters were not under civil service in those days, and the signature of a politician of the party in power was required on the application for employment. Furthermore, occasional contributions would be expected during political campaigns. This did not deter Joe. Although he did not approve of the sponsorship and donation requirements, he was motivated by his overriding personal mission in forestry. Joe completed the application and promptly walked over to the Senate chambers, where his good friend Senator Taylor endorsed it.

CHAPTER 10
Pennsylvania State Forest Management Plans

Several events had occurred previously that opened up that ideal opportunity for Ibberson in Pennsylvania in 1948. A 1940 article by H. G. Matoon in *Forest Leaves,* the magazine published by the Pennsylvania Forestry Association, set the scene. Though acknowledging that the Department of Forests and Waters had accomplished much through fire protection and production of seedlings, the article decried the lack of forest management:

> Even though several thousand acres have been in state forests for as long as forty years, no management plan has ever been developed and put into force. In recent years the excuse for delayed adoption of a system of management has been that the stock survey has not been completed. . . . It seems incredible that 1,650,000 acres of forest land . . . should be held and administered year after year in a haphazard manner. . . .
>
> The Department can and should be severely criticized for neglect of this, one of the most important of its activities.
>
> There is a general belief that the employees of the Department of Forests and Waters do not know or are not interested in the marketing of the wood products . . .
>
> The fault lies in an inadequate organization and a disinclination by the Department to grasp its responsibilities. It is stupid to preach planting and protection and do nothing about utilization.

Then in 1947, Governor James Duff decided to cancel all timber-harvesting contracts from state lands until suitable forest management plans could be drawn up. During World War II, 75 million board feet of timber had been harvested from state forests without regard to proper forest management. In those timber sales, all merchantable trees that were over a specified diameter limit had been cut. Citizens had been complaining about horrible logging on state lands and the unsightly debris. The cancellations of logging contracts with the Commonwealth generated lawsuits by loggers and sawmill owners, who were understandably upset.

Governor Duff was an outdoorsman, very forestry-minded and a promoter of clean streams. Besides mandating better forest management, he also proclaimed that 50 million seedlings were to be grown in state nurseries by 1950. His directive exerted great pressure on the forest districts to increase tree planting. For example, in one district, 2.2 million seedlings were planted within two years on 2,200 acres of coal-mining spoil banks, with the help of foresters temporarily transferred from other districts.

Before and even after World War II, the work of the twenty-four forest districts had been largely custodial, with no active silvicultural management. Each district forester, aided by one or two staff foresters, worked mainly on state park operations, forest fire protection, maintenance of forest roads and buildings, and tree plantations. Parks were becoming increasingly popular, so the foresters were spending more time on them. Governor Duff realized that timber management plans were essential to guide sustainable harvesting of trees. Therefore, accurate, geographically based information was needed on timber volumes, quality, and growth rates as a basis for management plans. Haphazard inventory data gathered in the 1930s, though extensive, were not designed for formulating management plans, nor were the data suitable for this purpose.

That was the situation in 1948 when Joe reported for duty to James Morton, chief of the Division of Forest Land Management, in the Education Building on the State Capitol grounds. Ibberson had become an employee of the Bureau of Forestry in the Pennsylvania Department of Forests and Waters. Joe's office was on the fourth floor, in a very good forestry library. Several dignitaries had occupied this office previously, including Joseph Illick, a former state forester from 1927 to 1931 who in 1948 became dean of the New York State College of Forestry in Syracuse.

Joe in his office, room 404 in the Education Building in Harrisburg. From here, he directed all the programs that he started.

Al Mickalitis, who had a master's degree from Syracuse, occupied the same office and was extremely helpful to the newcomer. Al had been spending a considerable amount of time with Governor Duff doing forestry work near the governor's mansion at Indiantown Gap. Al's friendship with the governor was often helpful during Joe's early days with the Bureau of Forestry. Later, when Ibberson became division chief, he made Al his trusted assistant.

Joe had become interested in forest management plans during his studies at Penn State and had been urged to pursue them through his political connections. Now, after Joe had honed his skills at Penn State and Yale, the mission for which he had been preparing practically since childhood was about to begin.

Now Ibberson was placed in charge of developing the first forest management plans for 2 million acres of Pennsylvania state forests. This was the opportunity Joe had dreamed about, and he was able to put to good use all the knowledge he had absorbed as a student. Joe immediately started writing a manual of procedure on how to formulate forest management plans for the twenty-four forest districts (reduced to twenty districts in 1952). The manual, which consisted of fifty-four pages plus thirty-four pages of appendixes, described in extensive detail the nine phases in conducting forest management inventories:

1. Preliminary procedure
2. Field reconnaissance
3. Growth and condition study
4. Maps
5. Analysis of field plot data
6. Aerial photo interpretation
7. Compilation of cruise data
8. Compilation of category descriptions
9. Subsequent inventory procedures

The huge project was to guide the mapping and inventorying of every forest stand by forest type and site class. During the eighteen months that it took to complete the manual, Ibberson's friend Miles Ferree at the SUNY College of Forestry was most helpful in reviewing the manuscript and making suggestions. Professor H. Arthur Meyer assisted in designing the procedures for selecting sample plots and doing statistical analyses.

However, the start of actual fieldwork would have to await the approval of O. Ben Gipple, who as the state forester was also director of the Bureau of Forestry. Gipple had been promoted to state forester in 1947 with the support of Harvey

Taylor. Extensive inventories and growth studies were needed as a sound basis for the management plans. But no approval by State Forester Gipple was immediately forthcoming.

In the meantime, Morton gave Ibberson some other assignments. The thousands of acres of tree plantations on state forestlands were to be thinned for the first time. The plantations had become overcrowded, so that growth rates were very slow and the weakened trees had become more susceptible to breakage and windthrow. The thinning project turned out to be successful, but only after district foresters were educated and convinced about the need. They were all older men, and at first they didn't want to cut down any of the trees they had planted. Yet it was a good opportunity for Joe to get to know the field personnel and become familiar with the state forests.

Joe also was assigned the task of supervising the revegetation of spoil banks where coal had been strip-mined. This assignment was challenging and rewarding as well. Research was done to determine which tree species would grow best on the various overburdens being strip-mined from each major vein. The pH varied from 3.5 to 7.0, depending on the vein being mined. Ibberson and Tom Balsinger wrote a planting guide based on that research. The survival rate and growth rates of planted seedlings increased remarkably when the guide was followed.

Another assignment in 1949 used Ibberson's skills in a scientific study of Cook Forest. The purpose of the study was to save this cherished remnant of that primeval forest from the destructive impacts of visitors and natural forces. The 6,328-acre Cook Forest State Park in northwestern Pennsylvania annually attracted some 250,000 visitors from far and wide. This marvelous forest included some of the state's last few remaining groves of virgin white pines and hemlocks. Some visitors have called it the most sacred forest in Pennsylvania, as these towering giants create a spiritual, cathedral effect that nourishes the soul. Generations of the Cook family have worked to preserve this treasure.

John Cook (1788–1858), the first permanent American settler in the area, arrived in 1826. He saw the potential of the Clarion River as a water route, bought some timberlands, and founded the village of Cooksburg. There he built a one-story cabin and the first of many water-driven sawmills. John's son Anthony Cook (1824–1891) acquired more extensive timber holdings. Unlike other sawmillers, he loved and preserved the biggest and best trees, reserving 3,000 acres around his home. His son Anthony Wayne Cook (1855–1935) expanded the sawmill business to West Virginia, Maryland, Oregon, and Washington, cutting as much as 1 million board feet of lumber daily. He became one of the first to support the preservation of the magnificent redwood forests.

By 1910, Anthony Wayne Cook decided the great trees at Cooksburg should be preserved as a public park. For the next eighteen years, his determination and perseverance overcame many seemingly insurmountable obstacles. The Cook Forest Association was formed in 1926 to save the few areas of surviving old-growth timber. Endorsed by national natural resource groups and Governor Gifford Pinchot, the association raised $200,000 to help the Commonwealth purchase 6,055 acres from A. Cook Sons Company in 1928 for $650,000. Cook Forest became the first Pennsylvania state park that was acquired to preserve a natural landmark.

In the next two generations, Anthony Wayne Cook Jr. and his son Anthony Eaton Cook worked diligently to ensure that the sylvan giants of Cook Forest State Park would be protected. Anthony Eaton Cook, as author, artist, and naturalist, has been an energetic advocate for the preservation of old-growth forests of Central America and North America. Through many talks, articles, and extensive travels, he has raised awareness about issues affecting the future of these forests. His landmark book, *The Cook Forest, an Island in Time,* contains remarkable historical photographs, artistically beautiful color photos, and a cogent history of the most significant remaining old-growth white pine and hemlock forest. The book also tells about the incredible struggle of the Cook family and conservationists in the early 1900s to preserve the remnant Cook Forest, a struggle that helped inspire the creation of some of our national parks. Its Forest Cathedral area was designated a national natural landmark in 1969 by the National Park Service.

Back in 1949, the trees in Cook Forest were deteriorating at an alarming rate as a result of excessive trampling of soil and roots by visitors, because trails had not been constructed properly. Destructive agencies such as insects, diseases, lightning strikes, windthrow, and old age also were causing much damage. There was widespread public concern over what was happening to this irreplaceable forest, and the Cook Forest Association was concerned as well. This led the Department of Forests and Waters to agree to undertake a scientific study of causes and possible remedies. Public Law 295, which in 1927 provided for the acquisition and management of the Cook Forest Park, empowered the department to make and carry out such plans, including the profitable utilization of down and dead timber.

Ten professional foresters under the leadership of Joseph Ibberson were assigned to make this study. While living at Cook Forest, they inventoried all timber on the property, mapped all the forest stands, and prepared recommendations for alleviating destructive conditions. Their twenty-page report included a plan of action to halt the devastating deterioration of the white pine and hemlock monarchs. The plan recommended light thinnings to improve stand quality, fire protection measures, preservation of wildlife refuge trees, and employing a well-qualified forester as park superintendent.

A letter to Governor James H. Duff, written in June 1949 by Ibberson and signed by M. F. Draemel as secretary of the Department of Forests and Waters, requested approval of all recommendations in the report, including the plan of action. Governor Duff immediately approved the report, and the plan of action was followed until 1953. Plans for work projects were coordinated with the Cook Forest Association. The system of roads and trails was expanded to relieve crowded conditions in certain areas. A sawmill established on the site removed 14 million board feet of dead and down timber as part of the restoration activities. In 1951, Professor H. H. Chapman of Yale, an eminent authority on forest management, examined the Cook Forest operations. His report concluded: "The cutting that I observed has been correct in every particular, and should result in greatly improved health and better growth for the larger trees, which were the ones that were left."

A member of the Penn State varsity football team, Richard H. Rauch, even wrote a poem and sent it with a note saying, "For Joe Ibberson, with best regards and a prayer for a more peaceful future":

COOK FOREST

This is the forest primeval
The home of the borer and weevil
Where virginal trees lean thirty degrees
And shiver and shake in each murmuring breeze.

This is the forest protected
Much too long its virtue respected
By those with a flair for sentiment rare
It sheltered grandpappy, O woodman forbear!

This is the forest of yore
When the Pilgrims first came to our shore
These giants of old, their vast years untold
Are being destroyed by foresters bold.

And now comes a pliant regarding
The methods of Ibberson, Lewis and Harding
So the voice of the sage on the Post-Gazette page
Views with alarm and sputters with rage.

So prop up the doddering hemlock
And shore up the leaning old pine
Don't starve the poor borer and weevil
Let them live as in auld lang syne.

Joe began another fascinating project while he was getting started on the management plans. Producing the booklet *Common Trees of Pennsylvania* in 1952 was one of his most enjoyable and rewarding projects. Ibberson recognized that forest property owners could be encouraged to better manage and care for their trees if they could identify the species, learn some basic facts about them, and thus enjoy their beauty and utility. The booklet also could be used by students and organizations such as the Boy Scouts to give them an understanding of forestry and conservation.

Realizing that such a booklet could be very beneficial to the conservation and management of trees throughout Pennsylvania, in both rural areas and communities, Ibberson set out to produce it. He had some highly qualified staff members to help, including Albert B. Mickalitis, John E. Aughanbaugh, Caleb L. Morris, and the skilled artist James M. Francis. Ibberson called them to a meeting to discuss the purpose and contents, and then they determined the guidelines and worked together on the project.

The foreword and introduction explain how trees affect the lives of everyone, every day. Trees produce wood for home construction, furniture, and paper products. Their beauty and shade provide solace and inspiration. They help purify the air we breathe and the water we drink. They enrich the soil, prevent erosion, and provide food and shelter for wildlife.

The next pages define and illustrate the terminology of leaves, twigs, fruit, and other tree parts, followed by fifty-four pages describing each of the more common trees of Pennsylvania. Drawings of the leaves, twigs, and fruit depict identifying features. Francis made all sketches from fresh, natural specimens. Descriptive information in a standard format provides additional details, including uses of the tree. To ensure that the descriptions would be easily understood, they were kept simple and short, focusing on the key characteristics for identification and avoiding technical terms wherever possible.

The team of five foresters sensed that they were undertaking a very important and far-reaching assignment. They really enjoyed working together and developed a high esprit de corps. Their efforts produced a real classic.

Common Trees of Pennsylvania quickly became very popular and has remained in high demand, more so than any other state publication. The booklet has been reprinted every few years, at least eight times between 1952 and 1988. The initial size of six by nine inches was reduced in later versions to a pocket size of four and a half by six inches.

A new version was produced in 1995, with some differences from the original. Some species were omitted, others added, and changed scientific names corrected. The new drawings are different from those in the original version. Much of the text describing the leaves, twigs, fruit, and bark remains the same. Ten people are credited for the new version, but there is no mention of the five who produced the original classic. Yet the enduring popularity of the booklet pays tribute to their vision.

Meanwhile, working on the other projects and awaiting approval to proceed with the state forest inventories and management plans, Joe noticed that there was almost no communication between James Morton, his boss, and State Forester Gipple. He learned that Ben Gipple, who was a 1915 Penn State graduate and the first drum major of the Nittany Lion Blue Band, had wanted to work for the Bureau of Forestry immediately after graduation. But Gipple had been kept from state employment until 1943, when he became chief of the Division of Forest Management in the Bureau of Forestry. He blamed graduates of the Pennsylvania State Forest School at Mont Alto for denying him state employment.

Practically all district foresters and division chiefs were Mont Alto graduates, including Morton, even as late as 1948. And many were still resentful that the forestry educational programs at Mont Alto and Penn State had been merged in 1929. Penn State established the Department of Forestry belatedly in 1907, after J. T. Rothrock had given up on the recommendation for a forestry school that he had made to the School of Agriculture. And so in 1903, he had founded the State Forest Academy at Mont Alto. It was the only such institution created by a state, to train foresters to work in Pennsylvania's forest reservations, which evolved into the state forests.

With this proud history, the students, faculty, and alumni of Mont Alto were incensed that their alma mater was to lose its identity through the merger. They hung in effigy two officials of the Department of Forests and Waters, whom they thought were to blame. Most of the students transferred to other institutions rather than complete their degrees at Penn State. For several decades afterward, they harbored ill will toward Penn State graduates. Ben Gipple was not the only object of their lingering resentment.

They also were opposed to Ibberson, just a fresh guy and a Penn State graduate, coming in to change everything with his management plan project. They had become accustomed to mostly planting trees, fighting fires, and operating state parks, but they conducted no active silvicultural management. Ibberson visited every forest district to explain the management plan project, and gradually their opposition changed to support. By the time Governor Leader was elected in 1954, they claimed that their involvement was needed to implement the forest management plans.

The impasse that prevented the inventories and management plans from going forward was broken on the way to a meeting in 1949. Joe and some other staff members from Harrisburg attended a meeting in Huntingdon for the five regional foresters. In those days, the state was divided into five regions, with each regional forester being administratively responsible for about four forest districts. Joe and Ben Gipple happened to sit together during the train trip to Huntingdon. In the course of their conversation, Gipple asked what Joe was doing. He replied that he expected to work almost exclusively on forest management plans for the state forests and went on to explain in great detail how this would be accomplished. As the train pulled into the Huntingdon station, Gipple said, "You are just the S.O.B. I have been looking for."

On the ride back to Harrisburg, Gipple ordered Joe to proceed at full speed on the management plans. Joe was to answer only to Gipple and would have unlimited funds. Financial resources came from certain revenues during World War II that the legislature had dedicated for forestry purposes. Joe felt sorry that the project had been removed from the control of Morton, whom he regarded as a really dedicated and able forester.

Secretary of Forests and Waters Milo Draemel wanted the management plans to be completed as soon as possible, perhaps at the urging of Governor Duff. Joe immediately set about buying aerial photos for all state forestlands and a Saltzman projector for making compartment maps. He purchased Marchant calculators, modern at that time, for all the districts and signed a contract with IBM to process all data from the inventories and growth studies. He organized a staff in Harrisburg to supervise all field activities and compile the processed date from IBM.

Starting in 1950, Joe hired dozens of foresters to do the fieldwork, up to fifty altogether. Joe interviewed all the applicants, some of them several times, and was very selective in the hiring process. Later on, all of them advanced to more responsible positions, so that at times every district forester and all division chiefs in the Bureau of Forestry were people Ibberson had hired to work on forest management plans. Two of them even advanced to become state forester.

Normally, in those years when patronage was rampant, the approval of a county politician was required for hiring state employees, but Ibberson concluded that the requirement of a politician's signature on each application was not practical. Most of the foresters available for employment did not have the connections to get political endorsement and might not even try. His good friend Harvey Taylor came to the rescue, agreeing to sponsor and sign every application of a forester that Joe brought to him. Most of the applicants did not even know they had a sponsor. Later M. K. Goddard adopted a similar strategy. While director of the Department of Forestry at Penn State, Goddard had applications from forestry students sponsored in the same way by Sam Hayes, the state representative from Centre County.

During the next five years, these new foresters that Ibberson hired measured trees on thousands of ten-acre sample plots. They organized compartment maps and developed management plans for each of the twenty forest districts. Up to four foresters staffed every district to delineate forest stands and place sample

New Diameter Increment Rule

Various methods of determining and predicting growth rates require that the present d.b.h. and the d.b.h. ten years ago, both measurements outside bark, be determined.

The present d.b.h. outside bark can, of course, be measured quite accurately with a tape or calipers. To determine the d.b.h. outside bark ten years ago requires that an increment boring be taken. The width of the last ten rings multiplied by two is the last ten years' wood growth in diameter. For a given species, the bark growth during the same period is a certain percentage of the wood growth and it varies from one species to another. The numerical values of these percentages are best derived from the available ratio K of diameter outside bark D to diameter inside bark d. This ratio:

$$K = \frac{D}{d}$$

is known as the bark factor and it is easily seen that for a given K the bark thickness is obtained by the formula:

$$\text{Bark thickness} = \frac{1}{2} \quad (D\text{-}d)$$

$$= \frac{d}{2} \quad (K\text{-}1)$$

The bark factor is multiplied by the ten-year diameter growth of wood to arrive at the total ten-year diameter growth of the wood and bark. This ten-year growth is then deducted from the present d.b.h. outside bark to arrive at the d.b.h. outside bark ten years ago.

To eliminate the multiplications necessary under the above procedure and thus save time and reduce the number of errors, the writer designed a diameter increment rule. By measuring the width of the last ten annual rings on the appropriate bark factor scale the total ten-year diameter growth of wood and bark is read direct.

This is possible since, $I = 2r \, . \, K$ where I is the ten-year diameter growth of wood and bark, r is the width of the last ten annual rings, and K is the bark factor.

It was found through experience that scales able to measure four inches of diameter growth were adequate. To find the length of the scale for any one bark factor, the following formula was used: Diameter growth of 4 inches

$$= \frac{4}{2K} \; . \; 1$$

Each answer from the above formula is then divided by 40 to determine the scale graduations for each bark factor. For the species growing in Pennsylvania, scales for bark factors from 1.04 to 1.15 are adequate.

A supply of these rules was made by a commercial rule company for the Pennsylvania Department of Forests and Waters during the summer of 1953. During 1954, after the rules were used extensively on a major inventory and growth study project, they were found to be entirely satisfactory.

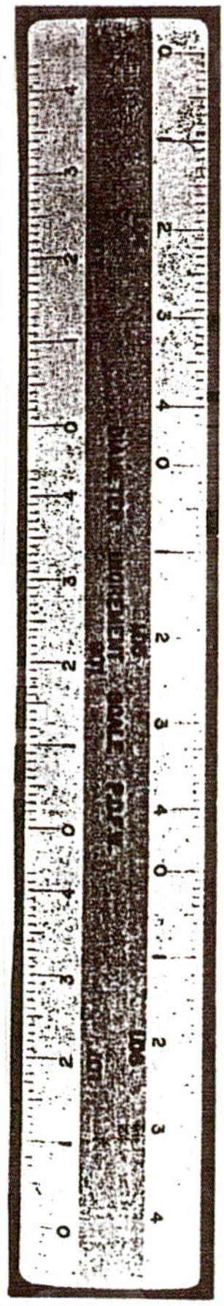

Be sure to give the SERIAL NUMBER as well as date of filing, name of applicant, and title of invention when inquiring about this application.

Address only
COMMISSIONER OF PATENTS
WASHINGTON, D. C.

DEPARTMENT OF COMMERCE
U. S. PATENT OFFICE

SERIAL NUMBER
529390
SERIES OF 1948

APPLICANT IBBERSON, JOSEPH EDGAR -- HARRISBURG PA

AUG 19 1955 (Filing date)

INVENTION DIAMETER INCREMENT SCALE

Case initially assigned to division 23

JOSEPH EDGAR IBBERSON
3764 SHARON ST
HARRISBURG PA

The petition, specification, oath, drawing, and first fee of THIRTY DOLLARS of your application for PATENT above identified received. 203392

Respectfully,
ROBERT C. WATSON
Commissioner of Patents.

CREDITS ON DRAFTS OR CHECKS ARE SUBJECT TO COLLECTION

PO-103a
3-11-47

16—51584-2 (SEE OTHER SIDE)

Article in the Journal of Forestry *and patent application describing a diameter increment scale invented by Ibberson.*

plots on thousands of aerial photographs, and then verify or adjust the forest stands on the ground.

They measured every tree above six inches in diameter in the plots and took borings from samples of trees to determine growth rates. To expedite ten-year diameter growth measurements, involving both wood and bark growth, Ibberson invented a diameter increment scale that calculated bark thickness ten years earlier. He published an article describing the new diameter increment rule in the 1955 *Journal of Forestry* and applied for a patent on his invention.

All of the measurements were recorded using mark sense cards. The foresters sent the cards to the central office in Harrisburg for analysis by IBM computers and compilation of data. A district representative placed stand boundaries on compartment maps. Jim Francis, who had good artistic abilities, drew the final maps.

Of paramount concern were the uniformity and accuracy of the data being collected and the delineation of timber size classes. That required proper supervision of the field crews in the twenty forest districts. Bill Corlett, a forester working on the management plan for District 3, was given this assignment and transferred to Harrisburg. In his new assignment, Corlett did an excellent job of keeping the fieldwork organized and training and supervising new employees. He worked closely with IBM to check that cards were marked and processed properly and printouts were adequate and on time. He also supervised the writing of the management plans and edited them to ensure uniformity and quality. Bill became one of Joe's most trusted associates as new initiatives arose.

An event in 1950 led to the realization that there was new, fresh leadership in the Bureau of Forestry. Ibberson presented a two-hour lecture to the bureau foresters at their annual meeting, held in Science Hall at Mont Alto. In his lecture, he proposed what should be done by the bureau and how to do it. He described how the forest management plans were to be completed by 1955 for all of the state forests, as well as the need to revitalize state nurseries, combat forest insects and diseases, provide advice to private forest owners, reclaim strip-mine spoils, and start a wood utilization advisory service. "We have the programs, the money, and the people to be the best in the nation," he said. At the conclusion, he received a long standing ovation. M. K. Goddard, who at that time was director of the Mont Alto campus, observed that the bureau was elated that it had found the new leadership it had been seeking.

Ibberson noticed that Ben Gipple, who was seated next to him in the front row together with Goddard, did not appear to care too much for the lecture. Gipple may have felt threatened by Ibberson's display of leadership and the acclaim he received after his presentation. In the weeks that followed, Ibberson resolved to watch himself to maintain a supportive relationship with his boss. Ibberson prepared much the same speech for Gipple to present to the Pennsylvania Forestry

Ibberson presented an inspirational lecture to Bureau of Forestry employees at Mont Alto in 1950, eliciting a standing ovation.

Association a few weeks later. The lecture turned out to be a blueprint for the many innovations that Ibberson would bring to the Bureau of Forestry.

The replacement of Admiral Milo F. Draemel with Samuel S. Lewis as Secretary of Forests and Waters caused some apprehension in 1952. But that soon subsided, as favorable reports about the management project came to the new secretary from the P. H. Glatfelter Company in York County. Glatfelter was a friend of Lewis's, and the favorable reports caused him to think about speeding up the project. Secretary Lewis asked Ibberson whether the project could be completed in half the time if the appropriations were doubled. Ibberson replied, "The plans are just like good wine; they should not be unduly rushed." Lewis grinned approvingly and became a close friend of Joe's.

Secretary Lewis turned out to be an outstanding administrator. Previously he had served Pennsylvania as lieutenant governor, Secretary of State, Secretary of the Interior, and postmaster of York. Ibberson felt that Lewis was the best administrator under whom he had served. Lewis obviously had great respect for Ibberson, as he took him on field inspections to discuss issues and seek his advice.

Another challenge to development of the forest management plans came from George H. Wirt, who had been the first technically trained forester employed by the Department of Forests, in 1901. Wirt also had been the first director of the Pennsylvania State Forest Academy at Mont Alto and had retired from the bureau in 1946. He arrived at Ibberson's office with fire in his eyes, demanding an explanation of what Ibberson was doing to the state forests. After all, he had devoted his entire career to these forests. Delighted to have this opportunity, Ibberson thoroughly explained the entire forest management plan project. When Joe was finished, Wirt patted him on the back and said, "Good boy, good boy," then slowly left, mulling over these new developments.

The concept of "forest management plans" changed to "forest resource plans" as the work progressed, for it had become apparent that timber was not the only forest resource to be managed. Attention to multiple uses of forests was to become prevalent in the forestry profession. The first forest resource plan, which would serve as a guide for all the other plans, was completed for the Michaux State Forest in 1951. Calvin F. Glattfelder, Robert Hyde, R. Bruce Shannon, and Robert H. Rumpf took on this task, and Ibberson supervised them on a weekly basis. All were Penn State forestry alumni except Hyde. They were headquartered in Science Hall at Mont Alto. This first plan resulted in such favorable attention to the whole program that the P. H. Glatfelter Company hired Glattfelder and Rumpf, the St. Regis Paper Company hired Hyde, and Shannon started a successful forestry consulting business.

Both Glattfelder and Rumpf ascended in the company to the position of vice president and general manager. Cal Glattfelder wrote to Joe in 2004:

> It was a great pleasure to see the *PA Forests* Winter Issue devoted to your magnificent record of forestry accomplishments in Pennsylvania!
>
> I was always impressed by your active leadership in the early days with the Department and I know I learned a lot from your wise counsel and experience.
>
> Many thanks for your wonderful sharing of your forestry expertise and your life with so many others!

The Michaux State Forest Resource Plan was experimental, and it stipulated selective cutting as the sole method of silvicultural treatment. It was amended in 1960 to permit diameter limit cutting for salvaging oak stands defoliated by gypsy moths or other destructive insects. In 1965, M. K. Goddard authorized even-aged management on all state forests, as research had shown that clear-cutting could significantly improve species composition and timber quality. Goddard's directive to district foresters, prepared by Ibberson, included cautions to ensure that clear-cutting methods on steep hillsides would not cause soil erosion or loss of water quality. Furthermore, no clear-cutting would be permitted along railroads where there were fire hazards; within 200 feet of picnic areas or campsites; or in wet, swampy areas or frost pockets. Discretion was mandated near trout streams, bodies of water, scenic highways, natural areas, major trails, and other scenic areas critical for public relations.

By 1955, the entire project was finished on schedule. Separate forest resource plans had been written for each of the twenty forest districts, based on hundreds of compartment maps and thousands of growth and inventory plots. An allowable timber harvest was set for each district, consisting of approximately half the annual growth in volume. Thus timber could be marked accordingly, and logging contracts could be awarded competitively. No other state in the nation had such comprehensive forest resource plans.

Title page of the first State Forest Resource Plan, signed in 1951 by Joseph E. Ibberson, State Forester O. Ben Gipple, and Secretary of Forests and Waters Milo Draemel.

The U.S. Forest Service was conducting forest inventories during the same time period, using somewhat different methods. Pennsylvania invested much greater effort in its inventories and management plans, according to cooperators in the Forest Service. Furthermore, the Bureau of Forestry publicized its plans to obtain public support, so as to avoid criticism. About 1962, Pennsylvania's inventory methods were modified to make them compatible with those of the Forest Service, and thus extended the usefulness of both. Bureau relationships also have been maintained with experts in Penn State's School of Forest Resources. These partnerships have promoted cohesion in forest resource planning and leadership in modernizing management plans.

The forest resource plans have been revised at fifteen-year intervals, and in some cases more frequently. Although the original plans dealt mainly with timber, they also pertained to other components and uses of the forests, such as wildlife, water, and recreation. When the plans were revised, expanded attention was given to other purposes besides timber, including wildlife, endangered species, primitive and wilderness areas, wetlands, and recreation. Various bureau employees assisted in updating and expanding the management plans; revising statistical aspects and forest resource plans continued after Ibberson retired in 1977. In 2004, the most recent statewide resource management plan was adopted to provide for sustainable timber management, low-density recreation, protection of water quality, and habitats for forest plants and animals. Public comments from many meetings were taken into consideration.

To implement the forest resource plans, Ibberson realized that tree volume tables would be needed for planning timber harvests. When it became evident that conventional form class tables were inaccurate for some species in northern Pennsylvania, Joe decided they would make their own. Professor H. Arthur Meyer at Penn State helped in originating and designing the volume table project in 1955. It was a cooperative project in which the Bureau of Forestry collected the data, the U.S. Forest Service carried out the calculations, and the Penn State School of Forest Resources compiled and published the volume tables.

Ibberson supervised the team that did the fieldwork. As one of the members described it, "Joe explained the job, turned us loose, and said to let him know if we had any problems." William Corlett and Thomas Delong made the initial measurements and developed guidelines for the project. By 1959, Corlett, Delong, Patrick Lantz, John Bitzer, William Vanidestine, and Jack Denniston had measured more than 4,000 trees. They sampled each commercial tree species in four geographic regions, on three site categories, and in each timber type. Penn State professors Ronald A. Bartoo and Russell J. Hutnik published boardfoot volume tables in 1962, and professors Peter Dress, F. Yates Borden, and Bartoo published cubic-foot volume tables in 1966. Professor Brian J. Turner published revised board-foot and cubic-foot volume tables in 1972.

CHAPTER 11

Organizing Pest Management

While the inventories and forest resource plans were taking shape, Ibberson sought other ways to improve the management of Pennsylvania's forests. He proved to be adept at getting administrative support and funding for several initiatives that expanded and improved the capabilities of the Bureau of Forestry. The most pressing of these was protecting forests against insects and diseases.

By 1950, gypsy moths were causing serious damage to Pennsylvania forests, and that was not the only problem. The oak wilt disease, Dutch elm disease, and other pests also were taking their toll. Gypsy moth infestations were spreading rapidly. Other defoliating insects periodically reached epidemic proportions. And Dutch elm disease, present in thirty-seven Pennsylvania counties, continued to kill hundreds of elms in towns and in the countryside. Ibberson had written an article about the dangers of oak wilt, and how to control the disease, even before it had been discovered in Pennsylvania. First found in 1950 in Mifflin County, the potential threat of oak wilt was being compared alarmingly to the earlier widespread destruction of chestnut blight.

Nobody at the state level was doing anything about these destructive organisms, so Ibberson decided to develop plans for a forest pest laboratory. Working with the Fisher Scientific Company, he drew up plans to develop a laboratory next to his office. He discussed his ideas with Ben Gipple, who took it to the Admiral for approval. As soon as Admiral Draemel signed his approval across drawings of the lab, Ibberson proceeded to have a completely equipped laboratory installed in the former State Education Building.

The first tree pathologist employed was Caleb Morris, a 1949 Penn State graduate in forestry. In 1951, Morris quickly cultured and identified the first oak wilt fungus found in Pennsylvania. Fifteen additional infected areas were found, and aerial surveys got under way in 1952. A survey and research program continued for some years in cooperation with the Pennsylvania Department of Agriculture, Pennsylvania State College, and the Game Commission.

The first entomologist hired was Ben Hadley, whom Joe recruited with the help of Dean George A. Garratt of Yale. Hadley published a paper in 1952 on lead arsenate sprays to control white pine weevil. Ibberson subsequently hired more entomologists, James Nichols, David Wallesz, John Quimby, Arnie Droos, and Ted Walker, and pathologists Barry Towers and Henry Gross.

Attempts to control the rapidly expanding gypsy moth infestations became a huge challenge. For years the gypsy moth had been confined to northeastern Pennsylvania, controlled there by spraying DDT, but the pesticide began to be phased out in 1958, and in the following years, gypsy moth infestations spread at an alarming rate. Ibberson started a laboratory at Middletown for rearing predators and parasites that might be able to control the gypsy moth. Dedicated by Governor Milton Shapp, the lab was regarded by the U.S. Forest Service as a leader in biological control methods. The Bureau of Forestry took on the responsibility for all forest insects because of a change in regulations. James Nichols transferred to the bureau from the Pennsylvania Department of Agriculture to handle the extensive work associated with this new responsibility.

In 1973, Secretary M. K. Goddard approved Ibberson's recommendation to State Forester Sam Cobb to form a separate Division of Forest Pest Management. James Nichols, who had studied the gypsy moth intensively and written an authoritative book on the insect, became its first chief. Millions of gypsy moth parasites and predators raised at the Middletown laboratory were released at infestations throughout Pennsylvania, and some of them eventually became effective. Barry Towers became division chief in 1986 after Nichols retired.

Frank C. Craighead worked on the oak wilt project under Joe's supervision in the 1950s. He was a 1912 Penn State graduate in forestry and a distinguished entomologist who had retired as head of Forest Insect Investigations in the U.S. Department of Agriculture. Insects transmitted the oak wilt disease from tree to tree, so his expertise was invaluable for the project. Frank had written a definitive 679-page book titled *Insect Enemies of Eastern Forests* (USDA Miscellaneous Publication No. 657), published in 1950. He also wrote a series of scientific articles on various forest insects during his career. He often said, "Study an insect and you are soon studying the whole forest. Study the forest and you are soon studying the insect."

Frank Sr. also was famous because of accomplishments of his children. He was the father of the Craighead twins, Frank Jr. and John, who achieved international acclaim for their pioneering research on grizzly bears and the National Geographic films about their work shown on PBS. His daughter Jean had been the first wife of John L. George, before he became a professor of wildlife management at Penn State's School of Forestry. An award-winning author whose books have sold millions of copies, she wrote popular books about wildlife for children. One book, *The Hole in a Tree*, was about a tree on a beautiful 120-acre farm near Carlisle that her father had bought for his retirement. She had observed the hole and its denizens for years and based the book on her observations. Once as she was showing Joe the hole in the tree, a cooperative fox squirrel stuck its head out.

Frank Craighead joined a research team studying oak wilt near Blain in Perry County, Pennsylvania. Several state and federal agencies were engaged in the extensive research project, as they feared the disease might become as destructive as chestnut blight. During the summer of 1952, Craighead and Caleb Morris observed more than twenty species of insects on the fruiting mats of the oak wilt fungus. They caged insects on wounds made on living oak trees and concluded that some of them could be vectors of the disease.

Ibberson and Dr. Tom Guyton, director of the Pennsylvania Bureau of Plant Industry, developed the procedures for finding trees infected by oak wilt. Observers in an airplane piloted by Bob Williams would spot the trees and mark locations on a map while they circled around. Then people on the ground would search for the trees and apply treatments to prevent the disease from spreading. After one flight, the pilot asked whether they were going to fly again the next day. Guyton replied that he did not want to insult his stomach again that soon; bouncing around on the air currents swirling over Pennsylvania's ridges and valleys had made him queasy.

While searching for oak wilt trees in Fulton County one day, walking some distance apart, Joe shouted out to Frank that he had spotted a very large rattlesnake. By the time Craighead arrived, the snake had crawled under a large boulder, but its tail was sticking out. Frank grabbed the snake by its tail, pulled it from under the rock, and flung it down hard on the ground. Then he pinned down the snake with his walking stick and grasped it behind the head with one hand and by the tail with the other.

Now this was a very large snake, and Craighead was a slightly built man. As the snake twisted and turned, Frank's body contorted violently in all directions. The snake appeared to be as strong as the man, and at times Ibberson thought the snake might be winning. Joe became very concerned as to what he might do if the elderly man got bitten, and the struggle was so scary that he himself might even have a heart attack.

After many anxious moments, Craighead somehow managed to kill the snake. Frank skinned the snake in a matter of minutes and announced that he would send it to a friend in Florida who was skilled at making belts from rattlesnake skins.

Frank Craighead at work in the forest. Photo provided by his daughter Jean Craighead George.

"Rattlesnake Craighead" became Frank's nickname. Ibberson just felt very fortunate that he and Craighead were unharmed, and able to walk down the mountain to their automobile.

A week later, Ibberson again was startled by a rattlesnake while looking for trees infected by oak wilt, this time in Perry County. Craighead and a third companion had already walked past without seeing the snake just three feet off the trail. Joe asked them to stop and gripped Frank's arm firmly with both hands. "What's wrong?" asked Craighead. Ibberson blurted out that they had just passed a rattlesnake, and he couldn't survive another experience like the previous week's in Fulton County. Their companion found the snake and killed it, while Craighead just looked on with a big grin. As Frank skinned the snake, he marveled at the beautiful yellow and black patterns and remarked that the old skin probably had been shed recently. He sent the skin to his friend in Florida and had a belt made for Joe, who still has the belt in his possession.

Upon recounting these two snake encounters to some older foresters, Ibberson found out why Craighead was so adept at handling rattlesnakes: Craighead and George Drake, both in the Penn State class of 1912, had paid their way through college by milking rattlesnakes. They sold the venom to a drug company in Philadelphia that used it to make antivenin. During summers, they lived in one side of a one-room cabin in Clinton County. On the other side, they kept dozens of rattlesnakes they had captured, separated only by a fence they had installed across one end of the room.

Joe and Frank became close friends and hunted and fished together. They visited M. K. Goddard several times in his office for discussions, at his request. Craighead was an astute woodsman and enjoyed the forests of Pennsylvania, but after some years of retirement on his farm near Carlisle, he decided it would be healthier and more pleasant to live where it is warm year-round, so he moved to Homestead, Florida. Joe spent several long winter vacations there with Frank, exploring the Everglades. Craighead had become an authority on that ecosystem.

Encounters with rattlesnakes continued at Homestead. One day, Craighead returned from the Everglades with a diamondback in the station wagon. He forgot to remove it or tell his wife about it. Imagine her reaction when she got into the car and heard the rattle of the diamondback! Another time, he picked up a huge rattler coiled in front of his daughter Jean in a hammock. On yet another occasion, he left a bag containing a rattlesnake in the attached garage. The snake escaped and surprised his wife in the bathroom. That could not have fostered matrimonial harmony either.

Craighead became a consultant for several influential environmental groups and moved to Naples, Florida, which was more convenient for these activities. He also wanted to be on the higher ground at Naples, having studied the destructive force of hurricanes and lived through Diane. He believed the Homestead area was due for another hurricane, and he was high and dry at Naples when the killer storm Andrew demolished Homestead. His command of all the disciplines involved in environmental issues, along with his ability to analyze and find solutions, made him very effective. Craighead tried to persuade Ibberson to become involved in the environmental groups, but Joe was too deeply rooted in Pennsylvania to consider such a change.

Upon reading these stories about her father, Jean Craighead George sent the following letter, dated July 6, 2005:

> Dear Dr. Gerhold,
>
> I have read with great pleasure and a lot of chuckles your account of Dad for the Joe Ibberson biography.
>
> I do remember Joe vividly as well as Dad's great interest in the oak wilt. He was so delighted to be working on another forest mystery and admired Joe greatly.
>
> I did not live with Dad and mother, but visited from Poughkeepsie, New York, where John George was teaching Conservation at Vassar College. I loved that farm and Dad's boundless knowledge about both plants and animals and their interrelationships. As he so often said, "Study an insect and you are soon studying the whole forest. Study the forest and you are soon studying the insect."
>
> Thank you for letting me read this. I added one little episode that shows Dad's uncanny insight into nature. He knew the Homestead area was due for another devastating hurricane and moved before Andrew hit. You don't have to use it as it may not fit, but I thought you would be tickled.
>
> And don't think for a minute that Dad was not in control of that rattler! He picked up a huge one coiled in front of me in a hammock in Florida.
>
> Jean Craighead George

CHAPTER 12

Revitalizing, Expanding, and Renaming the Research Division

As Ibberson's responsibilities expanded and the number of his employees grew, administrators in the Bureau of Forestry decided that it would be appropriate to accord division status to his group. However, creating a new division involved complicated procedures and required approval by the legislature. Then someone remembered that the bureau had a Research Division, which had been inactivated in 1938. The division originated as the Forest Research Institute, established at Mont Alto in 1930, the first research unit in the nation that was formed by a state. Henry E. Clepper and George S. Perry were senior research foresters, and in 1932, E. A. Ziegler became director of the institute. More than 250 research plots in natural forests and tree plantations had been started by the institute; John Augenbaugh, Tom Delong, and Ray Brooks had been measuring trees in this elaborate system while it was active. So the course of least resistance was to reactivate the Research Division in 1949 and name Ibberson as head. All of his employees were designated research foresters, even though they were not doing the usual kind of research.

The research foresters conducted various kinds of practical studies under Ibberson's direction. They experimented with fertilizers, insecticides, and fungicides to control white pine blight. They studied a disease of red pine and possible control methods. They tried out chemical and physical methods for controlling undesirable vegetation to improve the composition of second-growth forests and for eradicating Japanese honeysuckle from forests and weeds from nurseries. They compared planting methods for revegetating coal spoil banks, surveyed nut crops, and established eighty seed-collection areas. They studied problems at Cook Forest State Park and carried out recommendations for remedial treatments.

Ibberson took his expanded responsibilities as chief of the Research Division seriously. A substantial number of publications by him and his associates document their accomplishments. Between 1949 and 1955, he authored fourteen articles about research activities in *Pennsylvania Forests* and in *Pennsylvania Forests and Waters,* the department's magazine. His research foresters published thirteen more papers during these seven years, some jointly with Ibberson.

In 1956, the Research Division was renamed the Division of Forest Advisory Services. The name change resulted from a concern about the research responsibilities of the Bureau of Forestry versus those of the School of Forestry at Penn State. Representatives of Penn State decided that the proper place for research was at universities, whereas the role of a state forestry agency should be managerial. Actually, the Bureau of Forestry was just doing administrative studies to check on the practicality of research by others, as they applied to state lands. The bureau also provided land or facilities that enabled research projects to be conducted by other parties. As an example, the Bureau of Forestry provided some resources in support of research on oak wilt being conducted at Blaine by Penn State and the Forest Service.

Goddard decided that the Research Division would be renamed the Division of Forest Advisory Services. Ibberson had no enthusiasm for this change, but neither did he see anything to be gained by objecting to it. In fact, Joe wanted to dispose of the archaic forest study plots anyway, as they had become obsolete. So those records were transferred to Penn State, where faculty in the School of Forestry could bring them to a conclusion.

For Joe, nothing really changed. He did not let the change deter his aspiration to improve the state's forests. Instead, he continued with his forest management work, pursued other opportunities for advancing the effectiveness of the Bureau of Forestry, and easily adjusted to the mandate of the new Forest Advisory Services.

The new division was "to act as consultant for forestry operations on 1,800,000 acres of State Forests and promote better forestry practices on the remaining 13,200,000 acres of forest land within Pennsylvania." Under Ibberson's guidance, the Forest Advisory Services' staff grew over the years, as did their responsibilities. The staff expanded the management plans to address the preservation of endangered species and wetlands. They helped modernize the Bureau of Forestry through new programs by which service foresters advised forest landowners, and specialists in wood utilization and marketing distributed information to forest industries. Ibberson also started a forest genetics program and implemented improved nursery management practices that enabled the tree nurseries to produce large quantities of desirable and improved species.

Many of these staff members hired by Ibberson rose to positions of prominence in the Bureau of Forestry. One of them, Patrick Lantz, succeeded him as head of the Forest Advisory Services when Ibberson retired in 1977. In recognition of their creative and innovative abilities, advanced college degrees, and many accomplishments, they were dubbed the "Dream Team."

Secretary Goddard approved a plan in 1956 to hire eight service foresters for a Cooperative Forest Management program to assist private woodland owners by providing technical advice. Most other states in the region already had such programs, and actually, Pennsylvania's Bureau of Forestry also had been providing some types of assistance to private owners at least since 1946. Roy Olson of the Northeastern Area State and Private Forestry program of the U.S. Forest Service approached the bureau with a proposal. He offered to pay for the salaries and automobiles of eight service foresters for a year, but with a proviso: The Forest Service insisted that the service foresters had to be covered by civil service, and thereby may have helped secure such protection for all state-employed foresters. When M. K. Goddard became Secretary of the Department of Forests and Waters in 1955, he had persuaded Governor Leader to place all foresters under civil service provisions administratively. Then in 1956, the legislature made civil service permanent.

Ibberson thought the Forest Service offer was too good an opportunity to pass up, as service foresters really were needed. Most woodland owners did not have much technical knowledge about forestry, if they even thought about managing their forests. Goddard immediately agreed with Ibberson, even though the state forester, Ralph Wible, was hesitant about taking on more work.

To implement the new program, William Corlett was named chief of cooperative forestry; he held this position until he retired in 1985. He supervised the service foresters, trained them, and wrote annual reports to the Forest Service. Bill emphasized practical applications in the management of nonindustrial private forests and promoted a strong esprit de corps. He started annual educational meetings on cooperative forest management, which also were used by the bureau to train new employees. Bill frequently met with service foresters in the field, and sometimes they invited him home for dinner with their families. He knew every forester, his wife, and kids by their first names.

Ibberson worked closely with the new program. After 1962, several tree farms that Ibberson owned sometimes were used for training the service foresters. The tree farms served as practical examples of forestry methods and costs. There were twenty service foresters by then, and eventually as many as forty-five. The knowledge and experience that Ibberson gained from managing his tree farms was applied in advising nonindustrial private forestland owners. There are about 500,000 of these in the state, and they own more than 12 million acres.

Pennsylvania's service foresters program attracted regional and national attention. Ibberson and Corlett were known for being open to new techniques, such as log grading and tree grading, plotless cruising, and using the Bitterlich prism. Corlett occasionally was invited to Washington for policy discussions with various federal agencies. He also dealt with any problems of coordination, such as one with the Soil Conservation Service, an agency that wrote farm plans that often included woodlands. Its forestry recommendations sometimes were at odds with those of service foresters whom the farm owners contacted for assistance. Corlett and Ibberson arranged a memorandum of understanding between the two agencies, which solved the problem.

In 1955, Secretary Goddard stopped at Ibberson's office to inform Joe that he was to manage the forest tree nurseries. Ibberson remonstrated that he already had a full workload. But within the hour, he received a letter from Goddard making the transfer official and putting Ibberson in charge. New legislation that required the revegetation of mine spoils greatly increased the demand for tree seedlings from state nurseries. Tom Norris, a 1924 graduate of Mont Alto, had been administrator of the Nursery Section before Ibberson took on this responsibility.

Secretary of Forests and Waters M. K. Goddard, front left, signing an agreement with Ivan McKeever of the Soil Conservation Service. In the rear, left to right, are Joseph Ibberson, Vernal Miles, Ralph Wible, and William Corlett.

Ibberson promptly set out to inspect and evaluate each of the six state tree nurseries, and he spent much of his time on these trips. The Mont Alto Nursery had been established in 1902, and the one at Greenwood Furnace in 1906. He saw that the quality of seedlings needed to be improved, along with expanding production. The superintendents had not kept up with current methods and were using outmoded practices. Three months later, when Goddard asked him about the nursery program, Joe reported that all nurseries were using seedling production methods that had not kept up with advances in knowledge and were not very efficient. Therefore, Goddard decided to replace some of the nursery superintendents with professional people.

Thomas Delong headed a group of foresters who were newly hired as nursery superintendents. He reviewed the procedures of the six state tree nurseries to make them more efficient and improve the quality of the seedlings. Joseph Hill, who had been trained by E. J. Eliason in New York State nurseries, became superintendent at the Mont Alto Nursery. He assisted all nurseries in soil management, constructed a refrigerated vault for seed storage, and installed equipment for soil testing, seed testing, and seed extraction. Patrick Lantz, after serving as Hill's assistant for part of a year, became superintendent of Penn Nursery from 1960 to 1970.

Besides supervising the ordinary production of seedlings, Lantz greatly expanded and improved facilities for making signs and picnic tables for state forests and parks. He also installed and managed seventy-seven acres of seed orchards cooperatively with Jack Winieski when Jack took over the tree improvement program. Ibberson, Winieski, and Lantz together picked out the seed orchard sites and made arrangements for logging, burning brush, burying the stumps, and preparing the sites for planting grafted trees.

Ibberson was well aware of the importance of genetics for producing high-quality seedlings, even before 1953, when he attended the first of several Northeastern Forest Tree Improvement Conferences. Research and applications in forest genetics were in their infancy at that time. The first college course in forest genetics in the United States was taught at Yale in 1955; before then, just a few people had been conducting forest genetics research and starting practical tree improvement programs. In a 1950 article, Ibberson wrote that as early as 1927, the Department of Forests and Waters recognized the importance of using proper seed sources, having established its first quality seed orchard by converting a superior Scotch pine plantation.

In 1950, Ibberson hired Leroy D. "Whitey" Schaller, a 1947 forestry graduate of Penn State, who spent several years selecting superior forest stands and arranging for seed collections. Whitey had lived in the same rooming house with Ibberson when they were students and then went on to earn a master of science degree in horticulture. State Forester O. Ben Gipple had directed that additional seed orchards be established to meet the increasing demand for seedlings.

Schaller and Ibberson recognized that the strategy of commercial seed dealers was primarily to market large quantities of tree seed at low cost. Therefore, the seed seldom came from superior stands of trees and might not even be well adapted to Pennsylvania conditions. Joe facilitated the search for superior seed trees by asking district foresters to designate timber stands that might meet the requirements. Schaller developed a scheme that divided the state into climatic zones that would guide seed collections relative to planting locations. He examined 145 stands of thirty-four species and selected about 420 acres of seed collection areas. But because of health problems, Schaller was unable to continue this first attempt in Pennsylvania to apply genetics to improving the quality of trees grown in forest nurseries.

In 1962, Ibberson hired John A.Winieski as a forest geneticist to establish improved seed sources for use by the nurseries. He had met Jack, a graduate student at the SUNY College of Forestry, the year before at a Northeastern Forest Tree Improvement Conference. He also spoke with Ernie Schreiner, a pioneer forest geneticist working for the U.S. Forest Service, about his ideas for a tree improvement program in Pennsylvania. Bureau of Forestry employees Tom Delong and Joe Hill recommended Winieski after checking on his qualifications with his professors. Joe and Jack hit it off right away, as both were "coal crackers."

Jack selected superior trees of black cherry, European larch, Japanese larch, Norway spruce, and eastern white pine. He grafted scions from them into seed orchards located at Penn, Mont Alto, and Dague nurseries. Winieski also developed seed production areas for additional species, including white birch, white ash, black walnut, Austrian pine, bald cypress, and pitch pine–loblolly pine hybrids. Improved seed from the orchards was the first tree seed certified under the state's Department of Agriculture Certification regulations, which Winieski helped develop. Several of the seed orchards were certified, and about 45 percent of the seedlings grown in state nurseries were produced from orchard seeds in some years.

In 1957, Frank J. Schrey III was hired as the first wood utilization advisor in the Bureau of Forestry and assigned to Forest Advisory Services under the direction of Ibberson. Frank had graduated from Penn State in 1953 and received a master of forestry degree in wood utilization in 1957. Bill Corlett helped interview Schrey and subsequent wood-products employees, as they were placed under his supervision.

Schrey took over the *Pennsylvania Marketing Bulletin,* implementing major changes that made it a success. A marketing service to assist producers of forest products had been started in 1948. Through efficiency studies, Schrey provided technical advice to many industries. He created a *Sawmill Directory* and conducted a study of timber drain from Pennsylvania forests. He designed a large manufacturing facility at Penn Nursery, the one built by Patrick Lantz, for making picnic tables and signs for state forests and parks, greatly improving their durability and appearance.

John K. Sherwood Jr., a 1961 forestry graduate from Penn State, was hired to work with Schrey in assisting the forest products industry. Then in 1965, Ron Gale was hired to replace Schrey, who was about to move to the Department of Commerce. The industry was expanding rapidly, as were timber sales from state forests. These wood utilization experts conducted studies at many wood-using industries and recommended ways to make them more efficient and profitable. They also held technical workshops for log grading, lumber drying, and lumber grading.

After Schrey moved to the Department of Commerce in 1966, he continued to assist Gale and Sherwood in attracting new wood industries to Pennsylvania. Starting as a research analyst, he progressed to chief of the Research Division, then to economic development representative, and finally to manager of data services. The new wood industries included Proctor and Gamble's mill at Mehoopany, Masonite Corporation's hardboard factory at Towanda, a particle board plant at Mount Jewett, veneer mills at Sheffield and Hughesville, the Velda furniture plant in McElhattan, the Strick Truck Company's laminated trailer floor plant, and several others.

In the Bureau of Forestry and the professional organizations he joined, Ibberson had for the most part worked quietly behind the scenes, but effectively. He had attended many meetings of organizations, including the Society of American Foresters, the Pennsylvania Forestry Association, and for some years the Northeastern Forest Tree Improvement Conference. On those occasions, he often spoke with participants who were especially knowledgeable in some field of expertise or could be helpful to his forestry work in various ways. He commonly provided advice to leaders of the organizations, but he held few positions of leadership himself. Nevertheless, his colleagues and superiors were well aware of his expertise and accomplishments and appreciated them.

As a manager, Joseph Ibberson was a bright, self-confident, quiet, private person, and he was persistent in seeking practical solutions to problems. His questions were probing, sometimes about matters that at first seemed obvious but upon reflection deserved further analysis. At professional meetings, he did not seek the limelight or socialize very much. Instead, he often sought out individuals for discussions related to his work.

"Joe was very astute," said one employee. "He always got to the crux of a problem quickly. Even when his technical knowledge of a subject was limited, he knew enough to ask the right questions and analyze information presented to him."

Ibberson was elected a Fellow by the Society of American Foresters in 2001.

Here is an example of Joe's discipline. One morning at eight-thirty a request came in by phone to find a control for Japanese honeysuckle in Valley Forge State Park. By ten-thirty plans had been made to set up 110 seasonally replicated test plots. Arrangements were completed with the park superintendent to view and select the test site, and Joe dictated a string of letters to acquire herbicides.

Ibberson had a laid-back managerial style and knew how to delegate responsibilities. He would assign a task with clear goals and occasionally check on progress, without getting enmeshed in details, but he did expect to be kept informed. As another employee said, "Joe was a model boss and a friend to each of the employees working out of the office. He was attentive and protective, a facilitator, and well aware of problems associated when pursuing brand new research where everything is unknown and no prescribed course of action is available."

CHAPTER 13

Political Intrigues

Ibberson had pursued lofty aspirations to become a forester and to contribute toward improvements in people's lives. But his idealistic goals and actions were tempered by the realization that political influence could be helpful and, in fact, was unavoidable in working for state agencies. Politicians did indeed play a vital role in Ibberson's career.

The first political influence in Ibberson's career occurred in 1937, when State Senator George Kunkle endorsed Ibberson's application to the Forest Ranger Program at Penn State. Two signatures were required, along with that of the high school principal. A friend had persuaded Senator Kunkle to sign. Some of the professors at Mont Alto were impressed that a senator had signed Ibberson's application for admission.

State Treasurer Ramsey Black had been instrumental in securing Ibberson's first professional employment at Clark's Valley Dam. Joe's sister Ruth and her husband, Harry Stutzman, were friends of Black's, and he set up an interview with Don Carpenter of the engineering firm that employed Ibberson. Carpenter's friendship led to Joe's introductions to Councilman DeHart of Harrisburg, who hired Ibberson to be in charge of the DeHart Dam and Reservoir, and introduced him to the politically powerful Senator M. Harvey Taylor.

Taylor's friendship was the key to Ibberson's being employed by the Department of Forests and Waters and to his support for many subsequent accomplishments. He also supplied the endorsements for the fifty foresters whom Ibberson hired. Ibberson was so close to Taylor that he addressed him as "Pop Pop," which was acceptable only from his closest friends. Taylor's support was vital in the days when foresters were not under civil service protection. Senator Taylor continued to support Ibberson's constructive conservation programs throughout his entire career. He lived to be 104 and was in fairly good health up to his last days.

The role of J. Hugh McNeil was more complex, as he was both a state employee and a political operative at times. In 1948, when Ibberson was hired by the Department of Forests and Waters, McNeil directed its publicity division and served as editor of its magazine, *Pennsylvania Forests and Waters.* They became acquainted almost immediately, because Ibberson's activities often were the subjects of articles in the magazine. They also became social friends, who lived close to each other in Harrisburg and at times commuted to work together.

McNeil also was part of the Taylor political machine. At Taylor's request, McNeil would take administrative leave to manage political campaigns. These included gubernatorial and senatorial campaigns, as well as Taylor's candidacy when he was running for reelection. As McNeil's guest, Ibberson attended many meetings of the political elite in such places as Philadelphia, Pittsburgh, and New York City.

At one of these meetings, Ibberson met Blain Hocker, who was a coal dealer from Steelton and also a member of the State House of Representatives. Hocker was a natural, homespun type of person. People often told him that he spoke and acted just like Andy Taylor of Mayberry in the popular television show. Hocker usually responded that he had been like Andy before Andy came along.

Hocker took a great interest in Ibberson's forestry programs, and they quickly became friends. Hocker referred to Ibberson as a "coal cracker" because his hometown of Lykens was within the coal region and he actually had mined coal for a short time. Ibberson reminded Hocker that it was the coal crackers who supplied the retail coal for his business, so he should respect them.

Before long, Hocker was made chairman of the Appropriations Committee, which was perhaps the most powerful committee of the state legislature. Things were really looking up now for Ibberson, who could count among his friends the two most powerful members of the state legislature, Blain Hocker and Harvey Taylor. They proved to be helpful repeatedly in legislative matters.

A series of events began unfolding two years after Ibberson joined the Department of Forests and Waters, through which Ibberson's career became entwined with that of Maurice K. Goddard. The episodes began five years before Goddard was appointed Secretary of the Forests and Waters. Subsequently, Ibberson was able to use his political connections on several occasions to secure support for Goddard's highly acclaimed initiatives. Thus Ibberson worked behind the scenes for some of Goddard's grand accomplishments in improving Pennsylvania's parks and forests.

Goddard might not have been appointed in 1955 had not the Pennsylvania State Police initiated an investigation in 1950. A preliminary report alleged that a large commercial tree nursery had purchased seedlings from the Bureau of Forestry at a low price, and then resold them illegally at a much higher price. As the police continued their investigation, State Forester O. Ben Gipple started a parallel investigation, believing it was his duty. Gipple assigned many employees to work on the investigation. He asked Ibberson to assist too, but Joe resisted, pointing out that he was a professional forester, not an investigator. The extent of his responsibilities was to formulate the first management plans for the state forests and organize technical support services.

By 1952, the allegations and investigations of the seedling purchases were beginning to take on political overtones, so the news media were showing considerable interest. Admiral Milo F. Draemel, the acting Secretary of Forests and Waters, was not confirmed by the Senate. He was replaced by Samuel S. Lewis, appointed by Republican Governor John S. Fine. The scuttlebutt was that the new Secretary was going to fire Gipple and all those assigned to the investigation, as Lewis suspected they were spies for the opposition party. Lewis was a strong political figure who exerted considerable influence with the governor. By the end of 1953, the forestry investigators, including Gipple, were no longer working for the department.

In 1954, the newspapers really went to work on the tree-seedling scandal, as it became a hot topic in that gubernatorial election year. Gipple switched his allegiance to the Democrats and kept the newspapers informed about the investigation. Early polls indicated that George Leader, the Democratic candidate for governor, had no chance of winning. But as the election drew near, the gap narrowed, and he finally won by a very slender margin. The tree-seedling case may have made the difference. Ben Gipple, the former state forester, thought he would be appointed Secretary of the Department of Forests and Waters as a reward for helping the Democratic Party, and many politicians thought so too. He had high hopes, but they turned to disappointment.

Governor-elect Leader appointed an advisory committee of three to assist him with appointments of cabinet members. The committee consisted of Herb Kohn, a lawyer from York County; Genevieve Blatt, the Secretary of Internal Affairs; and Patrick Kerwin, a lifelong friend of Ibberson's from Lykens. Ibberson was privy to inside information from the committee, as was his friend Tom Corbett, a deputy attorney general assigned to the Department of Forests and Waters. Kohn would inform his friend Corbett, and Kerwin would inform Ibberson about committee deliberations. Corbett and Ibberson then compared notes to assess the reliability and accuracy of the separate reports. Invariably the information about the committee's activities turned out to be consistent.

Gipple's name came up as a possible Secretary of Forests and Waters, but he was soon passed over. The committee had reservations and wanted additional candidates. That prompted Genevieve Blatt to comment that her brother may know of some possible candidates. Joe Blatt was a forester who had graduated from Penn State in 1943. He suggested that the committee should contact M. K. Goddard, head of the Department of Forestry at Penn State, as he would certainly be able to supply a list of candidates. Goddard met with the advisory committee, and they were more impressed with him than with any of those on the list.

As a result, Governor-elect Leader and Genevieve Blatt met with Goddard at a restaurant near the Democratic headquarters in Harrisburg. Upon meeting him, Leader was so taken with Goddard and his qualifications that he offered him the job. Democratic party leaders, though, were insistent that John Torquato, a long-time political boss of Cambria County, should be appointed. But Leader was determined to install a nonpolitical professional with a strong technical background, and that man was Goddard. Leader then called Penn State President Milton Eisenhower for permission to hire Goddard. Eisenhower replied, "You really know how to pick them."

Gipple's reaction to Goddard's appointment was that Ibberson must have been involved somehow, as he was a close friend of Goddard's. But Ibberson had not exerted any influence in the appointment process. Ibberson did call Goddard to congratulate him and offered to help in any way possible. Goddard's simple and direct reply was "I am counting on you."

When George Leader became governor, his appointments included members of his advisory committee and others whom Ibberson knew. Dave Rendell, a friend from Lykens, was chosen to be Leader's executive officer. Pat Kerwin became chairman of the Liquor Control Board. Herb Kohn was appointed to be attorney general, a position that became an elective office some years later. Joe now really had a foot in the governor's office.

Joe was offered the position of state forester when Gipple was fired, but he turned it down because he preferred to continue with his other responsibilities. Ralph C. Wible, who had graduated in forestry from Mont Alto in 1927, got the job instead. His father was a friend of Secretary Lewis's, which probably didn't hurt his chances. Wible's name was among the candidates recommended by Harry Griffith, the administrative officer of Secretary Goddard.

Goddard was pleased that Ibberson knew so many powerful politicians as friends and confidants. He frequently asked Ibberson to lobby in a quiet way for legislation in both the Senate and House of Representatives. Ibberson's efforts probably were very helpful, though it is difficult to gauge his effectiveness. Many of the conservation bills that Goddard supported were indeed passed, signed by the governor, and became law.

Goddard's first priority in his new position was to bring foresters employed by the state under civil service regulations. Governor Leader agreed, realizing that civil service protection was paramount in maintaining a top-flight professional organization. Leader lost patience when the legislature stalled on passing a civil service bill, so he issued an executive order on October 1, 1956, directing that foresters and some others would be employed under the direction of the Civil Service Commission. Those incumbents who could pass a rather difficult civil service test were to be retained. All new forester applicants had to pass the test before they could be employed from a list compiled by the commission. Some incumbents failed the test, but it was not a problem for any members of Ibberson's "Dream Team."

Secretary Goddard's directness and noble objectives soon won over the legislators, thus developing mutual respect and trust. Occasionally Goddard asked Ibberson to join him for lobbying sessions with senators and house members. Several times he directed Ibberson to lobby on his own, especially with Blain Hocker, the chairman of the all-powerful Appropriations Committee. Hocker and Ibberson remained very close friends with strong working relationships.

In 1955, when all appropriations were frozen and no new employees could be hired, Ibberson asked Hocker for funds to hire eight foresters, who were needed to help with timber sales under the state forest management plans that had just been completed. Ibberson explained that each forester hired would generate the equivalent of four times his salary in revenue from timber sales. Hocker agreed that this would be good business, so despite the freeze, he saw to it that the budget was increased to provide for the eight new foresters, and even asked whether more were needed.

During one meeting, Hocker complained that Goddard was entirely unwilling to engage in political bargaining. Ibberson jokingly replied that he himself was even less willing to take political considerations into account. Within a few minutes, Hocker agreed to Goddard's request.

M. K. Goddard sometimes called Joe "my forester." They often argued about issues but respected each other. Joe always was well prepared when he went into meetings with him. If Goddard disagreed about a proposed action, he would pace back and forth in his office "barking like a maniac" while Joe just sat there grinning. He knew that Goddard, recognizing the validity of Joe's position, would have to cave in. Then Goddard would sit down and say, "Go ahead, Joe, you're my forester." Joe once asked Harry Griffith, Goddard's aide, "Does he do this to anyone else?" Harry replied, "Only if he likes you."

Joseph Ibberson and M. K. Goddard at a forestry staff meeting in 1957.

Ibberson was especially eager to assist Goddard with the Oil and Gas Lease Fund Act, and he helped secure its passage by lobbying. Governor Leader signed it into law on December 15, 1955. Before then, all income from leases of oil and gas extracted from under state forestlands went into the general fund. Under the new law, all these revenues were placed in a special fund and restricted for use exclusively by the Department of Forests and Waters for conservation, recreation, dams, and flood control. Goddard determined the need for all such authorized projects and their locations.

The act was regarded by some as the most significant conservation legislation approved in Pennsylvania since the establishment of state forest reservations in 1895 through the efforts of J. T. Rothrock. From 1955 to 1970, about $20 million came into the oil and gas fund. The money was spent principally on state parks, as well as for the acquisition of 79,000 acres of interior holdings in state forests, which improved boundaries and management practices.

Goddard, along with Jack Rex, who was chief of the Division of Land Acquisition, and Bill Forrey, the director of the Bureau of Parks, spent an enormous amount of time and energy in 1963 lobbying the House and Senate for Project 70. The proposed $70 million bond issue required the Commonwealth's constitution to be amended. The money was intended for state parks, reservoirs, recreational facilities, and historical preservation. The statewide referendum was approved by only 113,109 votes out of 2.2 million that were cast. Ibberson helped out by asking his friends in the legislature for their support in passing Project 70.

To implement Project 70, Governor William Scranton called a special session of the legislature. Guidelines were set forth giving the executive branch the powers that were needed to proceed. The Commonwealth's eminent domain laws were revised, because many of the park and dam projects would require much land to be acquired through condemnation procedures. Of the $70 million, four-sevenths were allocated to the Department of Forests and Waters, two-sevenths went to local government units, and one-seventh was divided between the Game Commission and the Fish Commission.

The department held public meetings for every proposed park. Goddard, Jack Rex, and Bill Forrey described and explained the proposed project and answered questions. Members of the audiences often were contentious and acrimonious, especially about land condemnation. But Goddard and his team prevailed by taking on the hard questions, supplying frank answers and explaining how the new parks and projects would be advantageous. Within three years, the department had spent $60 million on 443 projects. Matching funds from local governments amounted to an additional $23 million.

Governor Scranton in December 1965 proposed a $500 million bond issue known as Project 500. Part of the funds would be used to develop the parkland that had been bought under Project 70. Goddard and his team lobbied relentlessly for Project 500. Testifying before the Senate, Goddard stated, "It is obvious to all that land acquisition alone cannot fulfill the outdoor needs of our people and that development of appropriate facilities is the next step toward the goals envisioned by Project 70."

A referendum in May 1966 was approved, and Governor Raymond P. Shafer signed the enabling act for Project 500 in January 1968. The Department of Forests and Waters and the Fish, Game, and Historical and Museum Commissions were allocated $125 million of the $500 million to develop recreational lands acquired under Project 70. Nearly all of these funds were spent within ten years, as required under the Land and Water Conservation and Reclamation Act.

The most bizarre political events that Ibberson ever witnessed were the establishment of a Department of Environmental Resources over Goddard's opposition, followed by the unimaginable appointment of Goddard as its first Secretary. With Ibberson's help in lobbying, Goddard fought this new department with every last ounce of energy he possessed. He was opposed to combining regulatory and managerial functions, including the absorption of the Department of Forests and Waters by the Department of Environmental Resources. And then to top it all off, he lobbied from 1979, when he retired as Secretary, until 1995 to finally succeed in splitting the Department of Conservation and Natural Resources from the Department of Environmental Resources, which then was renamed the Department of Environmental Protection.

Goddard lobbied the Senate against the new department day after day during 1968. Several times he asked Ibberson to accompany him, because of Joe's friendly relationships with Harvey Taylor and other legislators. Goddard's rough, gravelly, penetrating voice reverberated in the Senate chambers. He was so strongly opposed to the proposed department that he lobbied anyone who would listen, maybe even the page boy! But despite Goddard's best efforts, in 1969 Governor Shafer signed into law Act 275, establishing the Department of Environmental Resources.

The proposed Department of Environmental Resources would absorb and demote the Department of Forests and Waters and also include twelve bureaus from the Department of Mines and Mineral Industries; three from the Department of Health; the State Planning Board; the Division of Mines, Quarries, and Explosives from the Department of Labor; and the Soil and Water Conservation Commission from the Department of Agriculture. Some of the boards that would oversee and assist the new bureaus would include legislators and other politicians. That deeply concerned Goddard.

As Governor Milton Shapp took office in 1971, he appointed Goddard as acting Secretary of the Department of Environmental Resources. Then in April, he nominated him to that position permanently. For the next six months, a bitter and acrimonious battle raged throughout the state. The Sierra Club vented a long list of objections, including that Goddard had been adamantly opposed to the creation of this new department. Others claimed that Goddard was a political dictator. Finally the chairman of the Executive Nominating Committee was forced to bring the issue to the Senate floor. The vote was forty-six to four in favor of the appointment. Goddard immediately set about organizing this hodgepodge of bureaus into a workable department and he remained its Secretary until 1979. When Richard Thornburg became governor, he appointed Cliff Jones, a friend of Goddard's, as the new Secretary.

Eventually Goddard was proven to be right in his opposition to combining law enforcement duties with conservation responsibilities in a vast new department. All sorts of problems ensued. Soon after departing as Secretary, Goddard began lobbying for the breakup of the Department of Environmental Resources. In 1988, he submitted a report to the legislature explaining why the functions of resource management should be separated from environmental oversight and regulation. Funds for state parks and forestry had declined for years, while those for environmental regulators had increased substantially. Forced timber sales were needed to raise funds for the operation of forests and parks.

Goddard secured support from some legislators and conservation groups, and their help enabled him to prevail. He campaigned for Tom Ridge as governor. He persuaded Cliff Jones and Pete Duncan, two former Secretaries of the Department of Environmental Resources, to inform the legislature that inadequate attention had been given to the 114 state parks and 2 million acres of state forests. Ridge was elected, and soon after he took office, the House of Representatives passed an administrative bill in June 1995 that split the department into two. John C. Oliver, the executive secretary of the Western Pennsylvania Conservancy, became the first Secretary of the Department of Conservation and Natural Resources. Goddard won this, his last fight, but it took him twenty-seven years.

M. K. Goddard certainly was a remarkable man, with long-term vision and foresight. He and his team, through the Oil and Gas Fund, Project 70, and Project 500, had created a state parks system to benefit the greatest resource of all, the human resource. Goddard had complete confidence in his staff of foresters, so he was able to devote much of his time to his many other activities. Ibberson felt privileged that he had been able to assist Goddard with these remarkable accomplishments.

Goddard often discussed forest policy issues and forestry projects with Ibberson, and because Joe thought about these things all the time, it gave him an advantage in their exchange of ideas. During one discussion when things were going Ibberson's way, Goddard, his former professor, said, as he had before, "You are my forester; go ahead with your plans." Ibberson regards this as the best compliment he has ever received.

CHAPTER 14
Married Life

Though Ibberson was deeply absorbed in his professional work and engaged with his colleagues and men with political power, he also had a fulfilling home life. Joe had become acquainted with Evanka Tsenoff while both were students at Penn State in 1947. They met one fall afternoon at Old Main while signing up for a hayride. Soon after they had exchanged names, Joe began addressing her as Evon, pronounced like Yvonne. She was taking some courses at Penn State, having already graduated from Indiana University. Evanka also had attended Middlebury College in Vermont for two years. She spoke German, French, Spanish, and some other languages. Joe was immediately attracted by both her intellect and her beauty.

Joe and Evon kept in touch over the next several years, and Ibberson got to know her family. She was a football fan and occasionally attended games when Indiana played Purdue. Two of her brothers, Andre and Chico, were outdoor sportsmen and occasionally joined Joe for hunting, thoroughbred racing, and other sporting activities. Joe was especially impressed by Evon's father, Dr. Krestu Tsenoff. He had been appointed head of the Bulgarian Orthodox Church Mission in the United States and Canada by King Boris in 1922, with headquarters in Washing-

Joe met Evanka Tsenoff at Penn State in 1947.

The Reverend Dr. Krestu Tsenoff, Joe's father-in-law, while studying in Germany.

ton, D.C. After completing his higher education in Germany and Switzerland, he had been a professor at the theological seminary in Sofia, Bulgaria. He was also a bishop, called a protopresbyter, in the Eastern Orthodox Church and had built a church in Dearborn, Indiana. He received two medals from King Boris on separate occasions for remarkable service to humanity in his native country.

Evon's father wanted her to be educated and trained to work in the Bulgarian embassy in Germany or Washington, D.C., so she graduated with a major in German from Indiana University. Before any of the planned embassy work could materialize, however, World War II changed everything. King Boris III of Bulgaria, a close friend of Evon's father, died a mysterious death in 1943 after a challenging visit to Adolf Hitler. King Boris, though allied with the Axis, had refused to prosecute the Jews or declare war on Russia.

Joe and Evon were married in 1951 in a civil ceremony, as he was Protestant and she was Catholic. They bought the home east of Harrisburg, which has remained Joe's residence. One of Joe's former colleagues described Evon as a lovely, gracious lady who was an absolute gem. On several occasions, she modeled gowns for the Bal Masque. Another of Evon's talents was her artistry. She was a long-term member of the Harrisburg Art Association, and at public sales, her still-life oil paintings were purchased for use in doctor's offices or other displays.

Evon worked for the Pennsylvania Employees Retirement System for several years. Her extra income wasn't worth much to them after taxes, so she decided not to work any longer. She then became a full-time housewife, and her spare time was completely consumed by church activities and entertaining relatives, friends, and visitors from Bulgaria. Evon became active in the Masonic Order of the Eastern Star. Joe also was active in the Masons, and the 32nd Masonic Degree was conferred on him. He regarded the Masons as a group that mainly did good works; the rituals didn't interest him.

The couple had a second home for some years, to which they retreated on weekends and for summer vacations. Located high on a hill near Inglenook, it had a spacious balcony with a beautiful view overlooking the Susquehanna River. There Evon hosted many meetings of the Eastern Star, as well as family reunions. Evon loved this home and regretted that they had to sell it when the upkeep took too much time, after Joe started buying forestland and developing it as tree farms.

Every year, the couple took extended vacations of a month or longer together. Sometimes Joe took leave without pay when his absence exceeded his vacation time. They traveled extensively to Mexico (see Appendix B), and also visited Florida, Arizona, and California. They enjoyed learning more about the great outdoors, especially environmental and historical features. In Florida, they spent time with Frank Craighead Sr., who continued to correspond with Joe about his environmental interests and places they had visited together.

Occasionally Ibberson joined his wife for church services at the Holy Annunciation Macedonian-Bulgarian Church at Steelton. There the men sat to the right of the center aisle and the women to the left. The inspiring choir music always was excellent. On Easter Sunday, the whole congregation paraded around the outside of the church chanting, "Christ has risen." In the fascinating wedding ceremonies, the groom is crowned king with an ornate crown, and the bride agrees to be his loyal, faithful, and submissive slave in all ways!

The son of King Boris, Simeon II, visited the Ibbersons at Harrisburg in 1958 while in exile. He had succeeded his father as king under a regency and was forced into exile in 1944 when Russian forces occupied Bulgaria. At the time, he was attending the Valley Forge Military Academy and Junior College. After studying law and business administration, he became a successful businessman. The exiled king presented an autographed photograph to Evanka. Simeon returned to Bulgaria in 2001, founded a political party that won half of the parliamentary seats, and became premier. He instituted economic reforms and worked to secure Bulgaria's admission to NATO and the European Union. He was the only monarch in history to regain political power through a democratic election to a new official position.

All the Bulgarians whom Ibberson knew consistently claimed the "bragging rights" for Saints Cyril and Methodius. They also were boastful of Alexander the Great, as the original Macedonia included the present-day Greece, Bulgaria, and Macedonia. The brothers Methodius and Cyril were born to their Bulgarian mother in 826 and 827 at Thessalonica, Greece. They converted Slavs to Christianity through their invention of the Slavonic alphabet, which was composed of thirty-eight letters that were suitable for expressing Slavic sounds.

During social gatherings with the Ibbersons, their Bulgarian acquaintances sometimes displayed their sense of humor. Ibberson noticed that Bulgarians often joked about being tightwads and thrifty, or even miserly. Here are some examples that he recalled:

- Bulgarians buy those doughnuts that have the largest holes, because the holes are free.
- They turn off the lights when they sit around talking at night, because they can hear just as well in the dark.

Above: *Exile King Simeon II of Bulgaria gave this photo to the Ibbersons when he visited them in 1958.* Left: *Dr. Tsenoff's religious service cross, over 300 years old.*

Above left: *Liturgy Bible of Dr. Tsenoff decorated in twenty-four-karat gold.* Above right: *Medals given by King Boris to Dr. Tsenoff for humanitarian services to the citizens of Bulgaria.*

- They stop their clocks when not in use at night to keep the gears from wearing out.
- A Bulgarian lad once boasted that he saved money by running behind a bus instead of riding in it. His father scolded, "You could have saved much more by running behind a taxi."

Until the later years of his marriage, most of Joe's time was fully occupied with forest conservation activities. He worked on weekdays professionally as an employee of the Bureau of Forestry and in his spare time on tree farms that he was developing. Evon often entertained at the lakeside picnic areas Joe had constructed on his tree farms. They especially enjoyed the singing by church groups and others late into the evening. So Joe and Evanka, each with their own interests and activities, were well satisfied with their relationship.

But then Evon suffered a protracted illness from complications after a fall at home that broke her pelvis. She required round-the-clock attention for about ten years, during which time Joe's forestry activities nearly came to a halt. Sadly, Evon died from these complications in 1991.

CHAPTER 15

Ibberson's Tree Farms

Ibberson wanted to preserve some of the forestland he had fallen in love with as a boy, as one of his lifetime goals. His plan was to acquire considerable forestland and manage it so that he could eventually donate it for parks to be maintained by public agencies for the people of Pennsylvania.

As part of his strategy, Ibberson had several intermediate objectives. He wanted to combine smaller parcels into tree farms and overcome the undesirable effects of fragmentation. He intended to develop and operate these tree farms so they could serve as examples to other forest owners. He wanted to gain firsthand experience in owning and managing private forestland, so that he could advise others with confidence. His staff of twenty service foresters could disseminate such information to the half million owners of private forestland in Pennsylvania whose properties collectively totaled about 12 million acres. Ibberson anticipated that through timber harvests, the tree farms also would generate profits that he could use for philanthropic purposes. And he believed that operating the tree farms would be beneficial for his health, as they offered recreational opportunities such as hunting, fishing, and hiking, as well as picnicking or relaxing amid the beauties of nature.

Since 1962, Ibberson has purchased nearly 100 parcels totaling over 2,500 acres, some of which he sold again when warranted by circumstances. He started four separate tree farms, naming them Alpha, Beta, Stoney Valley, and Fishing Creek Valley. Alpha Tree Farm, the largest that he retained, consists of 450 acres. At age 89, Ibberson was still active in acquiring and managing properties, particularly parcels adjoining his tree farms and parks that he donated.

The acquisition of a forested property normally requires the services of a surveyor, a forester, and an attorney. Ibberson fulfilled all three functions in purchas-

Clifford Jones, former Secretary of the Department of Environmental Resources, and his wife, Carol.

Charles Strauss (left), director of Penn State's School of Forest Resources, and Ibberson Professor Harry Wiant.

Left to right: Robert Griffith, Ronald Ramsey, Larry Williamson, William Forrey, and Greg Schrum.

ing his properties. As a licensed, registered surveyor, he located property corners and surveyed boundary lines. As an experienced forester, he estimated the value of the land and its timber. Finally, he searched the title to the land and drew up deeds.

The ability to complete the entire process was an advantage not only financially, but also in dealing with prospective sellers. Before Ibberson contacted the owner of a property that interested him, he would locate the boundaries, estimate its value, and examine the deeds of the property and adjoining properties. Usually he knew more about the property than its owner did. As part of his offer, Ibberson would explain how he would manage and care for the land. The owner would be spared the cost and inconvenience of hiring a surveyor and a lawyer, as Ibberson could arrange for the survey and prepare the deed. The owner could simply sign the deed and receive the payment check. All of this was very attractive to the owner, so some of the transactions were completed the following day, although a few negotiations for key parcels took up to fifteen years. At times, there also were landowners who did not want to sell.

After purchasing a tract, Ibberson arranged for several kinds of management practices. Henry Hartman and his weekend surveying crew members, Bob Martin and Duke Hobaugh, would officially survey the property and paint boundary lines. Roads necessary for access and fire protection would be constructed immediately afterward. Altogether, twenty-six miles of first-class forest roads were constructed and nine ponds were built for wildlife, recreation, and aesthetic purposes.

Timber stand improvement work was carried out by a crew consisting mostly of foresters from the Department of Forests and Waters. They worked on week-

Bruce Corsnitz on his tractor with mower, used for building roads, installing game food plots, and controlling vegetation.

Above: *Ben Mentzer in his logging skidder.* Right: *Ben Mentzer with a load of logs destined for the sawmill.*

ends and holidays, and were given hunting privileges and free firewood, which they especially appreciated during the years when there was a national energy crisis. There always were more candidates than jobs, as they found the work challenging, educational, and enjoyable. Joe also worked along with them.

The crews worked on more than 1,000 acres, improving quality and species composition. For thinning operations, they felled culled trees and inferior species with chain saws, and removed grapevines from trees. They installed many wildlife food plots and planted more than 5,000 conifer seedlings, most of which wound up being eaten by deer.

At first timber sales were conducted by contract. Some of the timber harvesting was to salvage dead and dying trees caused by gypsy moth defoliation. The trees to be cut were marked by Ibberson and other foresters and sold under contract to the highest bidder. More than 8 million board feet of sawtimber and 6,000 cords of pulpwood have been harvested. Part of the proceeds was used to build roads and ponds and for stand improvement practices. Ibberson invested at least $200,000 derived from timber sales in common stocks and allocated the earnings toward philanthropic purposes.

Later Ben Mentzer did all of the logging. He is a careful logger, concerned about keeping up the tree farms' good appearance, and is considerate and completely honest. Ben is a real expert at selecting, felling, and skidding trees with

During the 1973 energy crisis, homeowners cleaned up every last limb on logging operations to be used for fuelwood.

Twelve miles of improved woods roads give ready access to the Alpha Tree Farm.

minimal damage to the remaining trees, and he keeps the forest vigorous and healthy. He does not high-grade, which is a detrimental practice of harvesting only the largest and best trees. He is adept at selecting high-risk trees that are declining in health, quality, and value. These are cut and marketed before their value declines further or the trees die. This leaves space for the high-quality trees

Two of the man-made ponds on the property.

that remain, which will then grow faster. He cuts the trees into logs so as to maximize their quality and sale price, and then drags the logs with his skidder to a landing, where they are loaded onto his logging truck with a crane.

Throughout the nearly three decades that Joe and Ben worked together, they applied certain forestry principles and business methods in the overall operation of Joe's tree farms. As a timber harvester, Ben relied on forest management directions from Joe to ensure that wooded areas were being used according to their nature for maximum benefits. These directions included aspects of dendrology, silviculture, forest protection, mensuration, engineering, and ultimately the utilization of timber.

For decades, Bruce Corsnitz has been building and maintaining the roads on the tree farms, as well as tending the game food plots. He owns and operates a bulldozer and a grader for road building. He is adept at determining locations for new roads and providing for all-important water drainage from the roads.

Water running down a road will do extensive damage by erosion, which increases with greater distances and steeper slopes. Bruce selects the surfacing material of shale or stone and arranges for its purchase. He is careful not to damage trees during road construction. To keep vegetation along the roads in check, Bruce sprays Round-Up herbicide. A skilled logger, Bruce also helps Ben Mentzer when necessary, and with his experience in the building trades, he willingly undertakes construction projects.

The Petroski family has regarded the tree farms as an important and enjoyable part of their lives for many years. Rick Petroski's wife, Betty, calls the Alpha Tree

Cindy Adams Dunn was named director of the Bureau of Recreation and Conservation in 2005.

Joe Ibberson with Harry Wiant, Ibberson Professor in Penn State's School of Forest Resources.

Farm their second home. About 1985, Rick started working at the tree farms, cutting firewood to be used for home heating. On the first day, Rick brought his young sons, Patrick and Joseph, along to play in the woods while he felled and bucked trees with his chain saw. When the boys were older, they helped their father with the firewood project. The three of them also worked at other tasks, such as tree improvement and pruning, repairing roads, cleaning leaves from drainage ditches, spraying by hand the vegetation that was crowding the roadsides, clearing debris from the roads, and removing unwanted vegetation from the ponds. They all enjoyed taking safaris on the tree farms' road systems to observe the abundant wildlife and check on the condition of the timber.

Son Joseph was so impressed with the forestry activities that he decided to become a forester, graduating from Penn State in 2002 at the top of his forestry class. As a student there, he worked for Professor Marc McDill and gained expertise in formulating forest management plans. Two years later, he returned as a graduate student to pursue a master's degree in forest resources management.

Above: *One of four picnic areas where visitors can pause for refreshments.* Right: *Young visitors are exploring the bear den.* Below: *Some of Joe's friends are "helping exercise the fish."*

When Ibberson needed to prepare a new management plan for the Alpha Tree Farm, he asked Joseph Petroski to write the plan. Paul Troutman, a veteran service forester, commented that it was by far the best forest resources management plan he had ever seen. Joseph was hired by the Department of Conservation and Natural Resources as a forester, but he continued to work on the Alpha Tree Farm in his spare time.

The assistance of all these loyal, dedicated employees and friends has been indispensable in operating and maintaining the tree farms, and Ibberson credits them as instrumental in his being named the 1998 Outstanding Tree Farmer by the American Tree Farm System.

Wildlife safaris are the principal recreational activity at the Alpha Tree Farm for guests and employees, and they can stop for lunch at any of the four picnic areas. Many kinds of wildlife are abundant along the twelve-mile road system. On a late afternoon, one may see dozens of deer, several flocks of turkeys, an occasional grouse, many gray squirrels and rabbits, and even red and gray foxes, bears, and bobcats. A large variety of songbirds and raptors also frequent the various habitats. Once Ibberson found a baby bobcat in the middle of the road. It was so young it could barely walk, and in fact looked slightly wet. But before Ibberson could decide what he might do with this cute kitten, it disappeared.

Left: *Bears fish regularly at the four ponds.* Below: *As many as five bears have been seen in one day at the Alpha Tree Farm.*

Many largemouth bass remain in the ponds, despite foraging by the bears, bald eagles, and ospreys.

The ponds are a magnet especially for bears and birds. Largemouth bass make waves along the shoreline, begging to be fed. Black bears are numerous and often can be observed trying to catch fish, with some success. They also roll over stones as they look for insects to eat, and on occasion as many as two dozen stones that bears had turned up had to be removed from the road. Bald eagles dive for the fish, and as many as three have been seen at one time. Ospreys and great blue herons also are observed "fishing without a license."

The only Pennsylvania buck on display at Cabela's store near Hamburg, Pennsylvania; Boone & Crockett score 168-6/8.

Trophy buck harvested at the Alpha Tree Farm. Top left: *By Joseph Petroski in 2004.* Top right: *By Scott Schultz, World Archery Champ, in 1993.* Bottom left: *By Ray Ibberson, Joe's nephew.* Bottom right: *By Carl Baumgardner.*

A large group touring the Alpha Tree Farm.

One of the goals of the Alpha Tree Farm is the sustainable production of large trophy bucks. Unregulated hunting would interfere with this goal, so the property is posted to prohibit trespassing and hunting by the public. Initially the tree farm had to be patrolled to enforce the No Trespassing signs. Interlopers soon got the message, after a few charges had been filed with the district justice.

Ibberson and invited guests, usually long-term employees, do hunt regularly at the tree farm. Deer hunters are instructed to adhere to strict rules. Rifle shots are to be aimed at the neck, preferably while the deer is stationary, to avoid crippling the animal. When a neck shot is successful, it drops a deer on the spot, preventing an injured deer from suffering for a long time. If more than one shot is taken, the shooter has to explain what happened. Bowhunters are required to practice extensively before the season opens and are expected to drop a deer with one arrow.

Joe telling Penn State students about the tree farm.

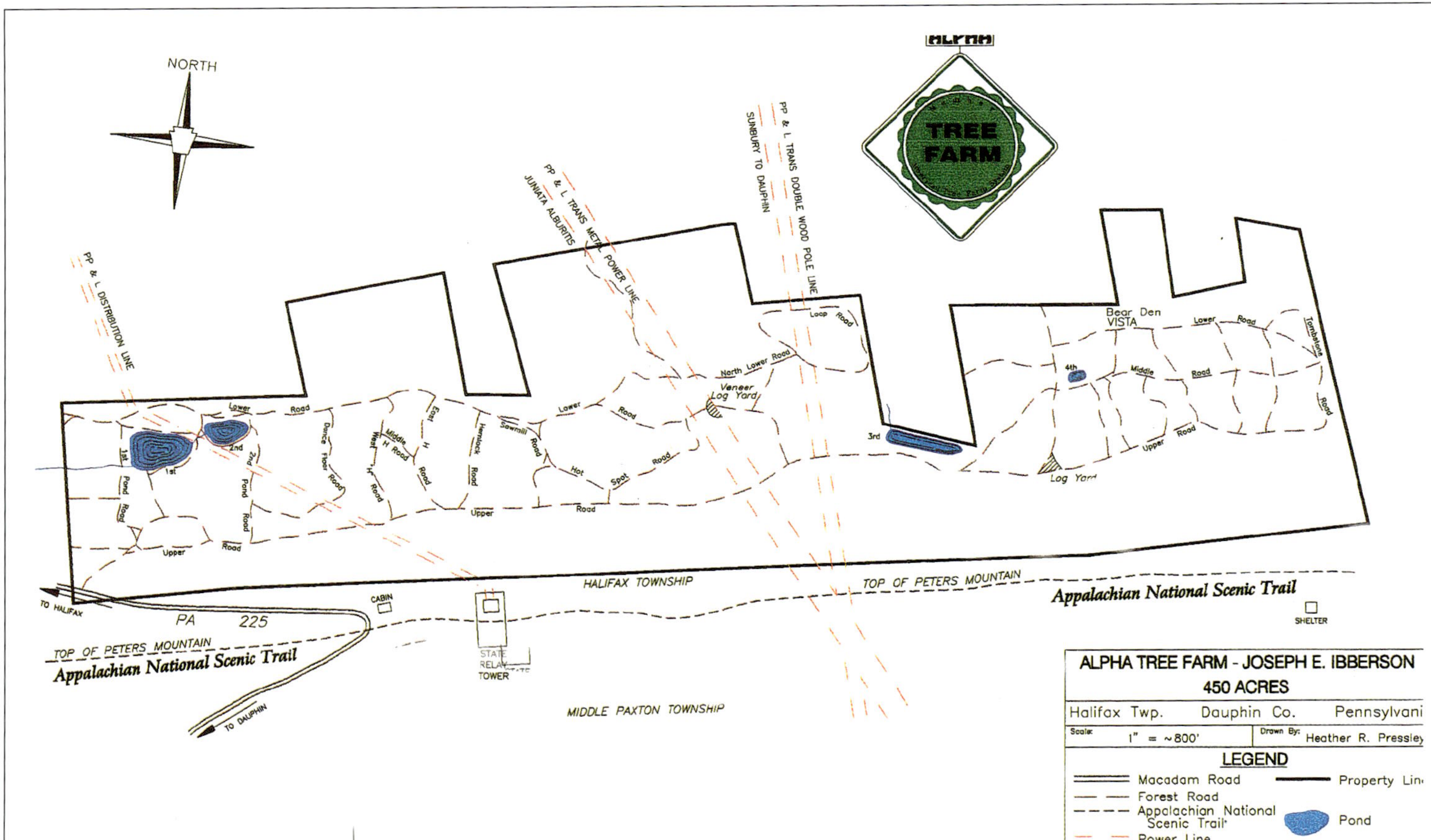

Map of the Alpha Tree Farm.

A working crew on the Alpha Tree Farm. Left to right: John Oliver, Secretary of the Department of Conservation and Natural Resources; Joseph Petroski; and Joe Ibberson.

Only bucks that have large racks with six or more points may be shot. Gradually, over a twenty-year period, the number of bucks with trophy racks has increased, until almost every buck taken has a trophy rack, with eight or more long points and a spread of at least eighteen inches. One day Ibberson shot a fine trophy buck that had ten points, and five minutes later he saw together two trophy bucks with much larger racks. In 2002, Joe's nephew, Bill Ibberson, bagged a twenty-pointer while hunting with him. He sold it for $3,500 to the new Cabela's store near Hamburg, Pennsylvania, the only buck from the state on display there.

The six-point rule is especially interesting in light of recently revised regulations of the Pennsylvania Game Commission. Upon the recommendation of Gary Alt, who promoted the state's new hunting regulations, which have been controversial, it is now illegal to shoot a buck with less than six points. The purpose is to increase the number of trophy bucks. Ibberson's experience of twenty years is direct proof that Alt's plan should work.

The Alpha Tree Farm (see Appendix C) has been used frequently for educating people about forestry and conservation. About six groups each year visit the tree farm for educational purposes. These include Penn State students, groups in the forest stewardship program, Penn State extension training sessions, and garden clubs. Observing accomplishments on the ground is especially effective and complements written information. The twelve-mile road system provides ready access to the various kinds of projects and practices that have been employed. In 1998, the Pennsylvania Forestry Association toured the Alpha Tree Farm at its annual meeting, attended by about 110 people, and its Tree Farm Committee presented the Pennsylvania Outstanding Tree Farmer of the Year award to Ibberson.

Ibberson has concluded that everyone should own a tree farm. When properly managed, a tree farm can serve as a gymnasium, chapel, social club, school, and bank.

CHAPTER 16
Financial Strategies

Ibberson started to accumulate quite some wealth after he turned thirty. Yet as a state employee, his salary was rather modest until he retired at age sixty. How did he manage to become so rich? Some have said he has a Midas touch, referring to the mythical king of Phrygia who was granted his wish to turn all that he touched into gold. That may be a colorful analogy, but in reality Ibberson's financial success can be attributed to his ability to make astute investment decisions along with some good luck.

Joe first became interested in the stock market while still a young boy because of comments his father made about people who had fared well financially. Joe observed that his father usually was right on the money. His father owned a few shares of the Lykens Hotel, which he had helped construct. The net worth of a fishing buddy, Swill Watkins, was over $1 million, derived mainly by investing in common stocks. His father also spoke about Dr. Nevin Baker, a dentist who became a multimillionaire and was the wealthiest man in Lykens. Ibberson recalled that during a visit to Dr. Baker's office when he was about twelve years old, the dentist left him in the dental chair several times to instruct his stockbroker to buy or sell some of his stock holdings.

In 1928, Joe's father took him to a stockbroker's office in Harrisburg, where the boy was fascinated by the ticker tape that recorded transactions in the stock market. He thought it was remarkable that the tape was reporting buying and selling activities so rapidly, and all the way from New York City. He also was impressed with the well-dressed customers sitting there watching the tape and occasionally placing buy or sell orders. Later, between 1940 and 1942, while employed by the City of Harrisburg, Ibberson traveled to the same broker's office and made several small purchases. During his employment by the Bureau of Forestry, Ibberson commonly frequented the stockbroker's office during lunch hours. He often invited forestry colleagues to accompany him, some of whom simply marveled at the ticker tape, while others were motivated to make investments themselves.

While still a youth, Joe learned some hard financial lessons. The stock market crashed in 1929, just a year after his first visit to the broker's office. Swill Watkins's million-dollar stock portfolio became almost worthless, but Dr. Baker managed to weather the storm and remained the richest man in Lykens.

Joe's personal misfortune in 1929—his loss of the first $100 he had earned by picking blueberries, delivering newspapers, and stocking store shelves—may have hit him as an even greater calamity. His savings account in the Miners Deposit Bank had been rendered worthless, as were the thirty-three shares of stock he had been issued. He had lost all the money he had saved for his education as a forester. Only three decades later did a silver lining appear on that cloud when in 1960, Ibberson sold his shares for $3,300 to a friend who wanted the stock so that he could become president of the bank.

While Ibberson was in the Army and stationed variously at San Francisco, New York City, New Orleans, Baltimore, Philadelphia, and Norfolk, he visited financial centers in his spare time. He observed, studied, and learned firsthand how stock markets functioned and behaved. The employees at those institutions were very helpful and friendly to this Army captain in uniform. Ibberson occasionally made small purchases at the stock markets in those six cities.

During his college education at Penn State and Yale, Ibberson took several economics and finance courses. This basic knowledge and after-class discussions with professors provided more information he could apply in making and growing his investments.

A $400 investment that Ibberson made in 1948 while at Yale could have paid for all the years he attended college. Fifty-six years later, it was still producing four fat dividend checks a year. A Mr. Lawrence from Boston demonstrated to a forestry class his truck-mounted Lawrence Mist Blower. This machine could deposit pesticides throughout entire tree crowns, especially the undersides of leaves, where insects could avoid sprays applied from above. Unlike conventional hydraulic sprayers, it did not create a mess or damage the paint of automobiles. Ibberson was impressed, and after the demonstration, he asked Mr. Lawrence whether any stock in the company was available. Upon learning that four shares were available at $100 each, Ibberson promptly purchased them.

This was the first phase of an investment that ballooned over the years. About a year later, a large tree care company bought the Lawrence Company, as the mist blower was creating competition for the tree company. For his four shares, Ibberson received a much larger number of shares in the tree care company, which also was paying quarterly dividends.

Several years later, the Elgin Sweeper Company, which manufactured street sweepers in Elgin, Illinois, bought the tree care company. Because both companies dealt with municipalities, efficiency could be gained by combining their products and services. Again Ibberson received a larger number of shares that had a greater value. The company had no debt of any kind, advance orders for two years, and a five-to-one ratio of assets to liabilities. It paid a good quarterly dividend as well as an annual stock dividend. Before long, the stock split two for one.

Next Federal Signal Corporation decided to buy Elgin Sweeper. Federal Signal was a large company that sold many kinds of products to municipalities—fire trucks, police cars, rescue vehicles, ambulances, garbage trucks, signs, and highway paints. Again this merger improved the efficiency of marketing. All of these indicators convinced Ibberson that he should retain his shares, which proved over time his astuteness as an investor.

Another of Ibberson's investments grew from $122,000 to more than $1 million. In 1989, he bought 2,000 shares of Smith Kline Beckman at $61 each. This pharmaceutical firm was headquartered in Philadelphia and had many promising drugs in the pipeline. The balance sheet met all his criteria, and the management team had a proven record. Furthermore, the price appeared to be low for the value represented by the company.

Soon Beechman Corporation, a large drug company in Great Britain, acquired Smith Kline Beckman and changed the name to Smith Kline Beechman. Part of the merger agreement was that Beckman Instruments would be spun off to the stockholders before the merger. Afterward there were other spin-offs, including Allergan, Baxter International, Beckman Coulter, and five lesser companies.

Then Glaxo Corporation, a giant drug company in England, acquired Smith Kline Beechman. After stock splits along the way and increases in shares resulting from the two mergers, in 2004 Ibberson owned 9,104 shares of Glaxo Smith Kline, valued at $426,000. In addition, several of the spin-offs had become more valuable than the stock of the parent companies. His Baxter International stock was worth $194,000, his Allergan stock was $477,000, and the stock from the other six spin-offs totaled $149,000. The grand total was $1,246,000, more than ten times the original investment. Dividends over the years have been substantial, too. In 2004, the annual dividends from all these companies amounted to $18,124.

Another investment was $50,000 in 1,000 shares of AT&T stock in 1960. The value of the stock of the parent AT&T as well as its many spin-offs, including Lucent Technology, ATT Wireless, and others, grew to about $400,000.

When the Weyerhaeuser Company went public in 1984, its stock was priced at $30 per share. Ibberson noted that the company had as many acres of timberlands as it had shares of common stock. Each $30 share thus represented the value of an acre of land plus a portion of all of the company's other assets. Recognizing this bargain, Ibberson acquired a sizable investment in Weyerhaeuser. Several years later, when the stock appeared to him to be overpriced, he sold it and realized a handsome profit.

Perhaps the largest profit Ibberson ever made in a short time started with a relatively small investment. While his father's will was being probated in 1964, he received a partial payment of $1,800. To honor his father, Ibberson decided to do his best to make this amount grow. He learned from a friend that 110 acres of forestland was available on Peters Mountain, a few miles from Lykens. The owner was a man who had retired to Florida. Ibberson located the property and thought it would be a good start for a larger tree farm. The elderly owner turned the matter over to his attorney, who had the property appraised. There was no legally recorded right-of way, so in an amazing coincidence, it was appraised at just $1,800.

Ibberson purchased the forest property and obtained a legal right-of-way. He had the property surveyed, the boundaries marked with white paint, and a road built for access and fire protection. Timber stand improvement practices were carried out wherever needed. But the prospects for adding land to make a worthwhile tree farm dimmed when a sawmill operator started buying most of the adjoining lands. So Ibberson decided to sell the property and start a tree farm at another location. After shopping around awhile, he found a buyer willing to pay $40,000 and sold the property to him.

Ibberson now had $40,000 to invest in something else. The United States at that time was on the gold standard, and the government was enforcing a price of $35 an ounce. He knew that international trade balances were settled with gold bullion, priced somewhat above $35 per ounce in the rest of the world. France and some other countries were settling their trade balances with the United States and taking our gold at $35. It was obvious that our country would have to abandon the legal price of $35 or have the vaults at Fort Knox emptied. Therefore, Ibberson invested the entire $40,000 in gold. Soon the United States abandoned the gold standard, causing the price to rise to the level in other countries. Immediately there was a significant profit in Ibberson's investment.

Then something unexpected happened. The Hunt brothers in Texas tried to corner the silver market by buying all available silver. The price of silver shot up from a few dollars to $80 an ounce. Concurrently, gold jumped to over $800 an ounce, as silver and gold prices usually move in unison. That price was laughable to Ibberson, so he sold his entire stash of gold. Thus the original investment of $1,800 from his father had increased to nearly $1 million. Certainly being lucky can amplify good judgment. The Hunt brothers were unsuccessful and took a major financial setback when silver fell to $5 and gold to $250 an ounce. Ibberson has avoided purchasing gold afterward. He invested the profits in forestland and stocks, whose values have continued to grow at a regular, steady rate.

Another investment Ibberson especially liked was bank stocks because, as bank robber Willie Sutton said, "That's where the money is." He was quite successful with investments in banks. Norristown National Bank was one of them. Bill Moll, the district forester at Norristown, pointed out that the bank's stock appeared to be a bargain. Ibberson checked it out, agreed with Moll's opinion, and bought several hundred shares. Before long, Meridian Bank of Reading bought Norristown National, integrating it with its sound, progressive management. Meridian paid large quarterly cash dividends plus an annual stock dividend. Stock also could be bought through the dividend reinvestment plan at a 10 percent discount, and an additional $5,000 worth of stock could be purchased quarterly at the discounted price. Ibberson took full advantage of these opportunities over a period of years, during which time there were several stock splits.

In the succeeding years, several mergers occurred. Meridian was taken over by Core State. The exchange rate appeared to be very favorable for Meridian stockholders, so Ibberson acquired a large number of Core State shares. Soon First Union Bank of North Carolina bought Core State, again at an excessive price. First Union lost some stature when it was accused of making some errors while absorbing Core State. It attempted to regain its good name by buying Wachovia, a southern bank that had an impeccable reputation. *Business Week* magazine in 2004 rated Wachovia the top bank in the country, and the University of Michigan Business School ranked Wachovia first in customer satisfaction. Several other banks unsuccessfully tried their best to buy Wachovia, so the competition was intense. Ibberson continued to invest via the dividend plan through all the mergers. Thus his initial investment in the Norristown bank, augmented by dividend reinvestments, eventually increased to more than $2.6 million.

Investment in the Dauphin Deposit Bank and Trust Company was another tale of success. This local Harrisburg bank was rated among the five safest in the nation, with fine employees who did everything right. Over the years, Ibberson accumulated common stock of Dauphin Deposit. When it merged with Allied Irish Bank, the stock exchange rate was favorable for Dauphin Deposit stockholders. In 1999, at a price of $121 per share, Ibberson's holdings in this bank were worth $2,420,000, and he decided to donate $1,500,000 to create and endow the Ibberson Chair in Forest Resources Management at Penn State.

Ibberson's sole reason for investing, after he had completed his college education, was for philanthropic purposes. Rather than being motivated by personal greed, as some investors are, Ibberson wanted to make positive contributions to humankind and thus make the world a better place. Perhaps that helps explain why success came to him so often and so easily.

To be successful as an investor, an individual needs to pursue a sound strategy of buying and selling and have a good understanding of his or her personal reactions to the vicissitudes of the marketplace. Timing is important, as is selecting stocks with good values and opportunities for growth. Having a clear purpose in mind and the ability to recognize reasonable gains related to the degree of risk are essential too.

Ibberson developed a strategy for investing in stocks that suited his own personality and philosophy. He added no more than one or two new selections of common stocks per year. Most of his purchases have been well-established, blue-chip companies that have desirable products, a favorable financial statement, and good management. He looks for the following useful indicators, though rarely are all these criteria met: The financial statement should show little long-term debt or none at all, as carrying charges are a drain on profits. The percent return on shareholder equity is a good measure of the degree of success of the company's operations. Cash flow is especially important, because that is the source of profits that can lead to growth in value; a larger cash flow is more likely to turn into larger profits. The ratio of current assets to current liabilities should be at least 2.0, indicating that the company can pay its bills and is doing so. The ratio of the stock price per share to earnings per share should not be higher than twelve; a lower ratio indicates that the stock is underpriced.

A good time to purchase common stocks of a well-established company is when its price has been beaten down for reasons that can be and are likely to be corrected. A recent example is the Hershey Foods Corporation. Its stock dropped to $35 a share when a new computer system that controlled inventories and deliveries was malfunctioning. There was fear that holiday orders would not be delivered on time and many customers would be lost. Stocks of this leading candy maker were being dumped for that reason. The computer problem was soon solved, whereupon the stock rose to $120 and afterward split two for one.

Ibberson has bought only common stocks and he has never bought any mutual fund shares, because he believes the small number of most promising common stocks he selects potentially will produce a much better return than the large number of stocks that most mutual funds own. He has never bought bonds, because they don't have as much a potential for large gains as common stocks. He has never bought any high-tech stocks, because he believed that their futures were too uncertain, and some that never have shown a profit had negative earnings and excessive prices. The few high-tech stocks that Ibberson does own came from spin-offs or mergers. In the 1990s, it became obvious that there was a bubble in stock prices that had to burst someday. When the bubble finally did burst, it had only a minor effect on Ibberson's stock portfolio.

Ibberson doesn't take advice from stockbrokers, whom he says are motivated by their own profits and may recommend buying or selling stocks just to generate commissions. Some brokers recommend buying stocks in which their employer has a special financial interest. Ibberson makes the stock selections himself and chooses brokers to make the trades who excel in executing the transactions.

Another of Ibberson's principles is to collect interest instead of paying interest. Paying interest reduces future purchasing power by depleting capital. Accordingly, it is judicious to postpone purchases until they are affordable. Then you can afford a lot more, because your purchasing power has not been reduced by interest payments. Although Ibberson has used credit cards for more than sixty-five years, he has never paid any interest or fees of any kind.

Ibberson also achieved great success in purchasing and managing forestland, totaling more than 2,500 acres since 1962. At that time, there was relatively little interest in mountain lands. Many of the owners paid taxes but had no income from the land, and some did not even know where their forestland was located. Ibberson bought some parcels for as little as $35 per acre.

As land and timber values grew over the years, the land Ibberson had bought increased in value to more than $3.6 million. In 2004, Ibberson received an offer of $2 million for his Alpha Tree Farm, which consists of twenty-one parcels purchased separately totaling 450 acres. Ibberson did incur the costs of building twelve miles of unpaved roads and four ponds, but these were more than offset by income from timber sales. The 350-acre Beta Tree Farm, consisting of eighteen parcels, was appraised at $1.1 million when Ibberson donated it to the state for a park. The Stoney Valley Tree Farm was estimated to be worth $225,000 when donated to Middle Paxton Township, where it is located, and the Fishing Creek Valley Tree Farm is worth $310,000. These property values plus about $400,000 realized from timber sales represent a total exceeding $4 million.

To calculate the profitability of the forestland investments, the purchase costs and management expenses have to be deducted. Expenses included land surveys, construction of twenty-five miles of roads and nine ponds, timber stand improvement practices, commissions to forestry consultants for timber sales, real estate and income taxes, and costs for travel, supplies, and equipment. Without going through all these calculations, Ibberson estimates these investments have gained about $3 million over a period of about forty years.

CHAPTER 17
Philanthropy

Joseph's philanthropic tendencies can be traced back to his mother and to his love of nature, which already was evident when he was a teenager. While exploring the mountains, valleys, and streams of Dauphin County, he dreamed that somehow he might be able to help preserve and protect these natural resources for the public to enjoy. That hope was reinforced after he graduated from Penn State, when he felt obligated to repay the university for all it had done for him. So as his wealth grew over the years, Joseph came up with ways of making his philanthropic dreams a reality.

By about 1995, Joseph decided that it would be advantageous to start implementing the provisions of his will while he was still living and could more fully enjoy the fruits of his labor. Not only could benefits to mankind be realized earlier, but he would be able to influence how his gifts would be used in a positive way. Thus the wisdom he could bring to bear, gained from a lifetime of practical experiences, could enhance the value of his donations.

Ibberson's first gift to an organization was a $50,000 endowment fund established for the Pennsylvania Forestry Association (PFA) in 1996, with the purpose of fostering the preservation and wise use of all forestland in Pennsylvania. Over the course of a year or more, he worked with several lawyers and a bank to set up the fund properly. The endowment fund had to be managed by a bank to meet federal regulations pertaining to such gifts, including the deductibility of additional gifts by others from their income tax returns.

The PFA is entitled to withdraw up to 5 percent of the fund's net asset value determined on January 1 each year. Any income above that is added to the mutual funds in which assets of the endowment are invested. By 2005, about $19,000

Linda Finley, president of the Pennsylvania Forestry Association, receiving a $50,000 check from Joe to start the Ibberson Endowment Fund.

Joseph Ibberson emphasizing the importance of forestry to Penn State President Graham Spanier, with Charles Strauss in agreement.

Left to right: Joseph Ibberson, President Graham Spanier, Dean Robert Steele, and Pennsylvania Forestry Association Secretary Norman Lacasse.

had been disbursed to the PFA, and the value of the fund still was somewhat greater than the original principal, despite large declines in stock market averages during some years.

The PFA has made excellent use of the income from the Ibberson Fund. In 2003, it established and supported a new Ibberson Forestry Forum in cooperation with Penn State's School of Forest Resources. The forum is designed to stimulate communication among forest landowners, wood-using industries, consultants, and public agencies.

Next Ibberson turned his attention to educating foresters at his alma mater, Penn State's School of Forest Resources. In 1998, he established and endowed the Joseph E. Ibberson Chair in Forest Resources Management with a gift of $1.5 million. Its purpose is to educate students and provide leadership in the professional management of privately owned forests. Ibberson's gift recognized that professional assistance is needed by more than 500,000 owners of private forestland in Pennsylvania, whose holdings total over 12 million acres. Undoubtedly that is where the management of forest resources has been overwhelmingly inadequate.

Some time passed while the agreement with Penn State was finalized, and before the right professor for the chair was found. When Charles H. Strauss became director of the School of Forest Resources in 2002, he promptly reorganized efforts to fill the chair. Previous searches had been unsuccessful, but this time, after interviews with several well-qualified applicants, Dr. Harry V. Wiant Jr. was selected and named professor of the Joseph E. Ibberson Chair in Forest Resources Management. Dr. Wiant had a nationally established reputation in this field, having previously served as president of the Society of American Foresters. He has been outstanding in molding young forestry students, devel-

Left to right: Ibberson Professor Harry Wiant, Joseph Ibberson, and Pennsylvania Forestry Association President Lloyd Casey at the first Ibberson Forestry Forum.

oping the first scholarly text on consulting forestry, and organizing the Ibberson Forestry Forum series.

Through his will, Ibberson committed a second gift of $2 million to Penn State in 2003, establishing a chair in urban and community forestry to provide leadership through instruction, research, and public service in the professional management of street trees, parks, greenbelts, watersheds, and forests of Pennsylvania's 2,675 municipalities. These urban forests touch the lives of all citizens in the state, so managing them properly is vital to all communities.

Through his will, Ibberson committed a third endowed chair to Penn State in 2006, to be devoted to silvicultural research dealing with the diversity of many organisms in complex forest ecosystems. He understood well that forests are dynamically shaped by climate, soils, insects, and diseases, and by societal demands and the managerial practices of forest owners. Ibberson wanted to promote silvicultural research that can improve ways to meet landowner objectives that are economically viable, ecologically sustainable, and politically acceptable.

In addition to making financial gifts, Joseph also wanted to influence how his forest properties could bring far-reaching benefits. Joseph valued spiritual enrichment just as strongly as the physical benefits of forests. His reverent feelings about nature and God have been pervasive throughout his life. He envisioned parks with unique qualities, where the usual kinds of active forest recreation are not permitted—special places where only passive recreation would occur, such as hiking, birdwatching, and nature studies. The overarching, cathedral effect of large trees would be preserved. No motorized equipment would be permitted except for park operations and maintenance. Physical and spiritual enhancement would be paramount.

Left to right: William Forrey, Larry Williamson, Joseph Ibberson, Gregory Schrum, and Ronald Ramsey.

Top: *The pavilion where groups can hear lectures and view the environmental surroundings.* Above: *The parking lot often is filled to capacity.* Left: *Restrooms and fountain near the pavilion.*

Joseph commented on the spiritual values of nature again shortly after the death of Pope John Paul II. He was reminded of stories he had heard of the Pope while he was a young priest near Krakow. There he used hiking in his ministries, especially with the youth. Because it was illegal under Communism for a priest to take students hiking, he would remove his collar and ask the hikers to call him uncle. During one such outing in 1967, he was informed that he had been appointed cardinal and was to return to Krakow. Throughout most of his years as Pope, he continued to enjoy hikes, and more than 115 were recorded.

To implement his benevolent intentions, Joseph contacted Roger Fickes, director of the Bureau of State Parks, and offered to donate his Beta Tree Farm. This 350-acre property is graced by mature stands of original forest, with trails for hiking, skiing, and other forms of recreation, and has many native wildlife species.

The pond is used extensively for studying aquatic life.

Nearby is the Shoop Site, where archaeologists have found thousands of artifacts, including Clovis points such as those associated with early settlers of Mexico. The spear heads indicate that Paleo-Indians had hunted migrating caribou and other animals in the vicinity more than 11,000 years ago.

Fickes and John Oliver, Secretary of the Department of Conservation and Natural Resources (DCNR), were delighted to hear of the offer. A long series of meetings ensued between Joseph and Gregory Schrum, chief of the Division of Resources Management and Planning. The two men decided that this new kind of park that Joseph envisioned was to be called a conservation area. But there were no legal provisions for the state to acquire and operate conservation areas with a restrictive covenant in the deed of conveyance. It took about two years to meet all the legal requirements for approval of a conservation statement of policy, defining a conservation area and its restrictive covenant, at all levels up to the governor's office.

Next a restrictive covenant had to be formulated that suited Joseph and the DCNR. The donor's restrictive covenant would express the purpose of the park and impose constraints on how it would be developed and managed. As they reached agreement on the various parts of the covenant, Schrum would obtain approval from Fickes and Oliver. Oliver had served many years as the executive director of the Western Pennsylvania Conservancy, and he took great pleasure in acquiring land for new parks.

In 1998, Joseph's gift of his 350-acre property in Dauphin County to the DCNR for the Joseph E. Ibberson Conservation Area became a reality. As an officially designated state park, it was the first of its kind in the state and probably in the nation. The restrictive covenant in the deed dedicates the land to the enjoyment and education of citizens of the commonwealth, emphasizing environmental learning and forest resources management. Only passive recreational activities

Department of Conservation and Natural Resources Definitions

What Is a Conservation Area?

This designation is for land donated to the Bureau of State Parks and managed for the purposes of preserving open space, conserving natural resources, and providing opportunities for passive, non-motorized, low density outdoor recreation and environmental education activities.

A conservation area is characterized as a large area with few improvements and no through roads. Recreational facilities and development are minimal. Conservation areas are used for low impact recreation and serve as outdoor classrooms. Conservation areas serve as examples of proper stewardship and resource management. The donation of land can be covered by a restrictive covenant.

What Is a Restrictive Covenant?

A restrictive covenant is a condition that is written into a deed, either by the seller or person donating the property, that must be adhered to by the person or organization that assumes possession of the property. Some of the restrictive covenants placed on the transfer of this property include:

- **The land is intended for use, enjoyment, and education of all citizens of the Commonwealth. Environmental, outdoor, and forest resources management education will be emphasized.**
- **Only passive recreational activities will be allowed. Horseback riding, biking, and motorized equipment, with the exception of official use equipment and accommodations for disabled people, are prohibited.**
- **Acceptable forestry practices will govern resource management. Generally, only dead and downed trees should be harvested, except on forest demonstration areas.**

are allowed. Timber harvesting is restricted to dead and downed trees, except for demonstration of acceptable forest practices. A wide variety of educational and interpretive programs are offered. The park has eleven miles of trails that Joseph had constructed previously for hiking and cross-country skiing, as well as a pond where an environmental education specialist teaches about aquatic organisms. In 2001, a pavilion and two restroom facilities were built, financed by the oil and gas fund that Ibberson had helped Secretary Goddard to establish.

Restrictive Covenant for the Joseph E. Ibberson Conservation Area

THE TRACTS ABOVE DESCRIBED BEING ALTOGETHER SUBJECT to any and all easements, rights of way and conditions which may affect the above described premises, and in particular those following COVENANTS and RESTRICTIONS which shall run with the above described premises in perpetuity:

a) The land is intended for the use, enjoyment and education of all citizens of the Commonwealth.

b) The land is intended as a place of exercise, contemplation, meditation and communing with the natural environment.

c) The land is to be a retreat from the pressures of their lives.

d) The land is to be a place to recreate and refresh the human soul. Only passive recreational activities will be allowed—horseback riding and motorized equipment, with the exception of official use equipment and accommodation for disabled people, will be prohibited.

e) The land is a place to learn how to manage resources for the resources' long-term health while providing for the maximum production of recreation, wildlife, water and forest products consistent with preservation of the environment.

f) The land is a place to be used as an outdoors environmental laboratory for research, learning and demonstrations, especially of proper forest management techniques for the private woodland owners of Pennsylvania. The Bureau of Forestry will assist with these activities and the Pennsylvania State University School of Forest Resources will be invited to participate.

g) All buildings and structures presently on the property will also be transferred but lifetime tenancy will be granted for use of the house trailer to the Grantor herein.

h) An important element of this forest will be areas set aside for big trees. Only dead and down trees will be harvested in these big tree set aside areas.

i) The foregoing restrictive covenants are permanent, irrevocable and of unlimited duration.

This conveyance is a donation of property to the Commonwealth of Pennsylvania and therefore is exempt from State and Local Realty Transfer Taxes.

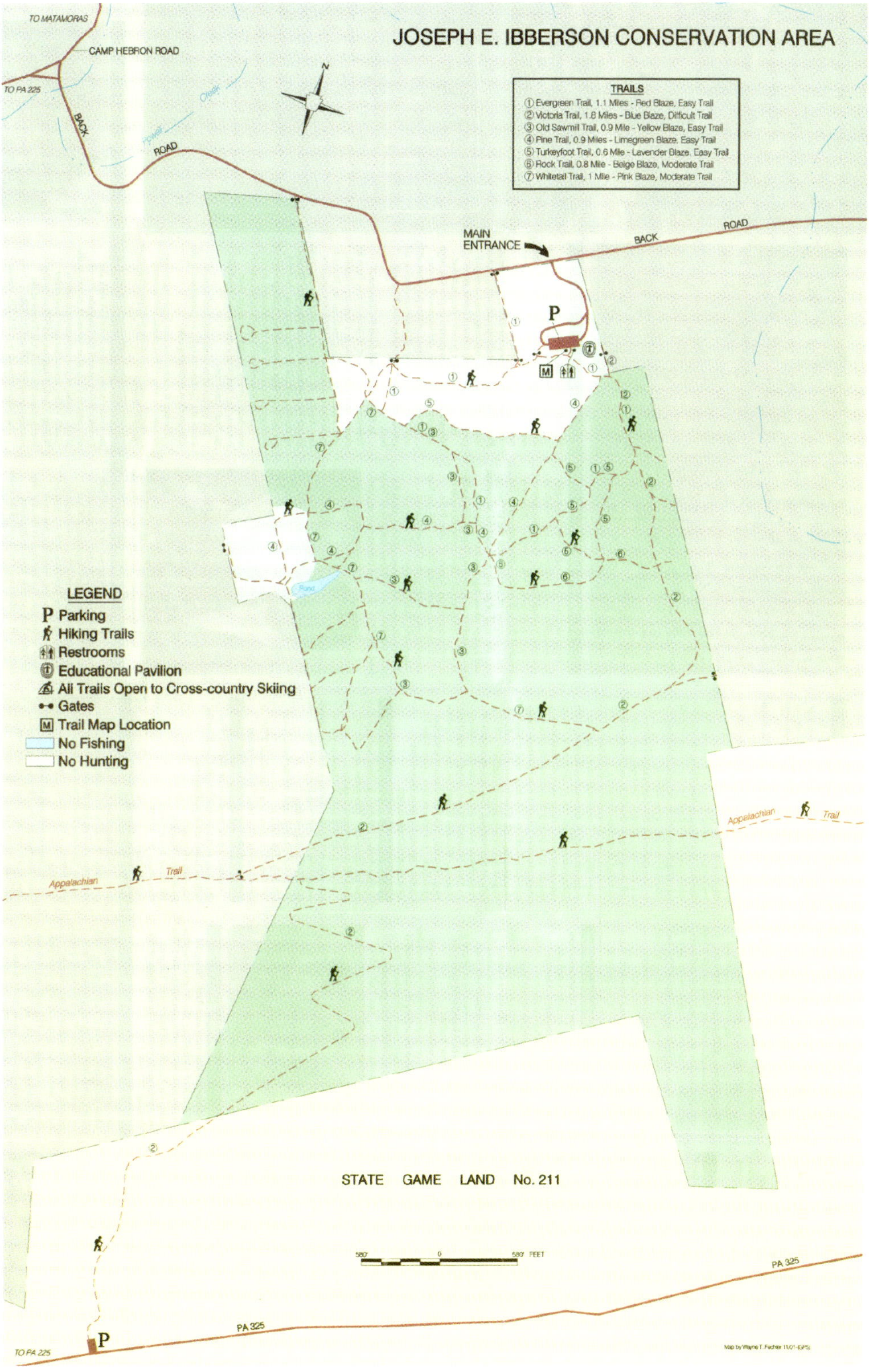

Map of the Joseph E. Ibberson Conservation Area.

Governor Tom Ridge issued a statewide news release to publicize this new park. Many newspapers and magazines spread the news, including a favorable editorial in the *Harrisburg Patriot News.* These articles generated a great deal of interest. Since the conservation area opened, many people have visited to experience its natural beauty.

The favorable publicity generated by the Ibberson Conservation Area stimulated others to consider making similar gifts. One was Joseph's friend Alex Boyd, president of the Union Deposit Corporation and a real estate investor, who wanted to explore the possibility of donating 900 acres for a conservation area. After a year of discussions involving Boyd, Schrum, and Ibberson, Boyd finally reached an agreement for a restrictive covenant with the DCNR.

In 1999, the 900 acres became a state park, named the Boyd Big Tree Conservation Area. The park is a forest jewel on both slopes of Blue Mountain, just a few miles north of Harrisburg. Twelve miles of trails wind through magnificent large trees of many species, across numerous wetlands, and along a clear, pristine stream containing native brook trout. These woodlands provide habitats for deep-forest birds, especially warblers. The many trails offer good opportunities to see white-tailed deer, wild turkeys, squirrels, grouse, black bears, and a variety of songbirds throughout the seasons. A large field filled with butterfly weed blooming in midsummer attracts many butterflies. American chestnuts were planted in the field near the entrance as part of a program to develop blight-resistant trees. The field and ridgetop of the conservation area can be great places to watch the annual hawk migration in the autumn as these wonderful birds of prey and other raptors ride the thermal currents along Blue Mountain.

The Boyd Conservation Area also received considerable publicity, resulting in even more donations of forestlands. Leonard Berman, another friend of Joseph's, called him to explore the possibility of donating a key twenty-four-acre tract

Girl Scouts starting a butterfly garden.

Above: *Environmental education specialist Stephanie Livelsberger telling mothers and children about the trees.* Right: *Mother and son studying lady's slipper flowers.*

adjoining Boyd's former property. After discussions with Schrum and Ibberson, Berman donated his addition to the Boyd Big Tree Conservation Area. The Central Pennsylvania Conservancy, through its president Carol Witzeman, purchased an additional eighty acres to provide a major corridor of access from the east.

Another gift of land was prompted by an article about the Ibberson Conservation Area in *Pennsylvania Forests,* the magazine of the Pennsylvania Forestry Association. After reading the article, Dr. Mead Shaffer, a veterinarian in Delaware County, called Joseph about 343 acres he owned in Wayne County. Shaffer, Schrum, and Ibberson spent a day exploring the property. They marveled at the magnificent natural forests and plantations, clear streams, wetlands, beautiful stone fences, and good road system.

Shaffer thought his property should be preserved because of its ecological value for environmental education, and after about two years of negotiations, he was finally convinced that a conservation area was the best of several options. He deeded the land to the state in 2001. Named the Varden Conservation Area for the adjoining village of Varden, it became the third conservation area of the Bureau of State Parks.

John Oliver, Secretary of the Department of Conservation and Natural Resources, played a decisive role in acquiring these three conservation areas. He told Joseph that they were the highlight of the eight years of his term in office.

The land and other donations of Alex Boyd, Leonard Berman, and Mead Shafferall inspired by Joseph Ibberson, together exceeded $11 million in value.

Above, left to right: *Department of Conservation and Natural Resources Secretary John Oliver, Alex Boyd, and Gregory Schrum discussed the creation of the Boyd Big Tree Conservation Area.* Left: *Alex Boyd in a large field of butterfly weeds, which attract thousands of butterflies.*

In 2004, Joseph donated forty-six acres to Middle Paxton Township for the Joseph E. Ibberson Middle Paxton Conservation Area, again with a restrictive covenant. It is believed to be the first conservation area with a restrictive covenant owned by any U.S. township. The main purposes of this conservation area are to serve as a retreat from the pressures of life, where people can commune with the natural environment, and to conserve the flora, fauna, wetlands, and forest.

The property is located three miles east of Dauphin, Pennsylvania, and is known as the Mary Bell Hollow. It is covered with a magnificent forest with springs, two ponds, small sparkling streams, wildflowers, and other rare flora. Gigantic trees are in abundance, including virgin timber consisting of oaks and tulip poplars, some over three feet in diameter. The stately trees create a cathedral effect that can enhance the visitor's feeling of well-being and spiritual awareness.

Ibberson also plans to donate an adjoining tract that he owns. The deed for this property will not include a restrictive covenant, as he feels the land is suitable for a ball field and a house there could serve as a Boy Scout field headquarters.

Joseph already has set into motion provisions that will facilitate conservation of more land for the public good. He was one of the key founders of the Pennsylvania Parks and Forests Foundation, an organization that had been a longtime ambition of Roger P. Fickes, director of the Bureau of State Parks. It took two years for Ibberson and other knowledgeable, experienced people to

Restrictive Covenant for the Joseph E. Ibberson Middle Paxton Conservation Area

FURTHER UNDER AND SUBJECT to any and all easements, rights of way and conditions which may affect the above described premises, and in particular the following COVENANTS and RESTRICTIONS which shall run with the above described premises in perpetuity:

a) The land is intended for the use, enjoyment and education of the public.

b) The land is intended as a place of exercise, contemplation, meditation and communing with the natural environment.

c) The land is to be a retreat from the pressures of life.

d) The land is to be a place to recreate and refresh the human soul. Only passive recreational activities such as hiking, walking, jogging, cross-country skiing, bird-watching and flora and fauna studies will be allowed. Horseback riding and motorized equipment, with the exception of official use equipment and accommodation for disabled people, will be prohibited. Hunting will be permitted for wildlife management purposes.

e) The many rare wetlands and their flora and fauna will be protected and developed.

f) Big trees and the flora and fauna found only in mature native forests will be of major importance and interest. Only dead and down trees will be harvested. The towering cathedral effect of the massive majestic trees remind our hearts to reach upward to heaven.

g) No land from this donation shall ever be sold, leased, exchanged or transferred.

h) The foregoing restrictive covenants are permanent, irrevocable and of unlimited duration, as long as streams flow and trees grow.

This conveyance is a donation of property to Middle Paxton Township, a Township of the Second Class of the Commonwealth of Pennsylvania and therefore is exempt from State and Local Realty Transfer Taxes.

Grantor hereby covenants and agrees that Grantor will warrant SPECIALLY the property hereby conveyed.

Leonard N. Berman
2525 Paxton Street
Harrisburg, PA 17111

October 12, 2004

I cannot really put into words all that I've learned from Joe Ibberson about forestry, conservation, and the proper care of land and the trees that grow on it.

Joe has cherished and protected our natural resources for more than half a century. Inspired by his interest, dedication, and commitment, many of Joe's friends have donated thousands of acres of unimproved land for green conservation, hiking trails, and glorious Pennsylvania woodlands. I am proud to say that through Joe, Phyllis and I donated 24 acres of land valued at more than $100,000.

Please join me in recognizing Joe's remarkable achievements to forestry in general, and, in particular, his immeasurable contributions to the conservation of Pennsylvania's beautiful trees.

Leonard N. Berman
Harrisburg

MEAD SHAFFER
VARDEN, PENNSYLVANIA

November 1, 2004

Dear Joe,

In 1999 I happened upon an article in Pennsylvania Forests Magazine describing your donation of land to the Department of Conservation and Natural Resources for use as a conservation area and environmental laboratory for research and education.

I had for some time been searching for just such an opportunity for my property in Wayne County.

Thanks to your help we now have set up The Varden Conservation Area which will soon open for public enjoyment, schools are already using the property for studies.

Without your inspiration and encouragement, this project might never have come to pass.

The donated property was appraised at $1,000,000.

Thank you again Joe.

Warmest regards,

Mead Shaffer

Alexander Boyd
P.O. Box 20730
Wickenburg, AZ 85358

Over time I had dozens of land options, but the most rewarding one I never exercised but gave to Joe Ibberson for 150 acres so he could complete an 800 acre tree farm on the second mountain in Middle Paxton Township, Pennsylvania.

From that day on I had a true friend whose knowledge of timberlands, timbering, road construction and land acquisition is boundless.

With his help I was able to acquire, develop and donate 900 acres as a state park named the Boyd Big Tree Conservation Area. It includes 12 miles of fabulous hiking trails and is located within a mile of Harrisburg. The appraised property value was over $8,000,000.

Without Joe's input the donation would never have come to pass.

In many ways Joe is one of my very best friends.

10/4/04

Coach Joe Paterno sending Joe Ibberson into the game of philanthropy, to continue his winning ways.

formulate a charter. Ibberson and Clifford L. Jones signed the incorporation papers to make the foundation a reality and launch it in 2000. The foundation has an executive director and an office in Harrisburg.

The Pennsylvania Parks and Forests Foundation helps with critical land acquisitions and financing for state park recreational facilities. Increasingly, individuals and organizations want to contribute money to a particular park or forest for some general or specific purpose, such as in memory of a loved one. As a tax-exempt organization, the foundation can accept such contributions and hold those dollars in escrow until proper arrangements are made to spend the donations as intended. As development and other activities encroach on parks and forestlands, the need grows to buffer our public lands. Land acquisitions include privately owned land inside a park and adjoining properties, which are ultimately transferred to the DCNR to maintain in the best interests of the public. The foundation also assists motivated citizens and businesses in forming partnerships and friends groups to support particular parks or forests.

Ibberson's decision to start his philanthropic contributions while he was still alive has given him a great deal of pleasure and satisfaction. Though somewhat amazed at how much has already been accomplished, he believes the best is yet to come. He has plans for himself and through his will to donate more parks, add to existing endowment funds, increase financial support for endowed chairs at Penn State University, and make other philanthropic gifts. He continues to search for new ideas to effectively preserve the natural resources of Pennsylvania and promote their wise use. His goal remains to make positive contributions to mankind and thus help make the world a better place.

Penn State Coach Joe Paterno spent some time with Ibberson in 2006 to discuss their common interests in philanthropy and sports. They enjoyed exchanging ideas and some good-natured kidding. Ibberson gave Paterno two books about the history of football, and Paterno presented Ibberson with an award from Penn State President Graham Spanier for his generosity and achievements.

CHAPTER 18
Retirement and Accolades

Ibberson retired from his employment by the state of Pennsylvania in 1977 at the age of sixty. Through his work in the Bureau of Forestry, his statewide accomplishments had met and even exceeded his lifetime goals of conserving forests. The Division of Forest Advisory Services was superbly staffed with talented foresters and functioned smoothly. Ibberson reasoned that his contributions could be enhanced in the long run if he retired and devoted all his time to managing his own lands and growing his financial assets. He also would need time to make fiduciary arrangements for gifts to worthy causes.

Joseph is certain that his mother, by reading to her children the Bible story of Joseph and his coat of many colors, had intended to challenge him to do good deeds for his family and for people in general. Like his biblical namesake, Joseph did perform good deeds for his family and his country. While he was gainfully employed, he helped his siblings and a nephew buy homes and automobiles and remodel their homes. He assisted some relatives in starting businesses and others in finding employment.

Besides helping family and friends, Joseph also provided leadership in government during his professional career. After retirement, he donated land and money through several projects that would benefit mankind and thus make the world a better place. Reflecting on his accomplishments he once commented, "I believe I was picked by the Lord to practice good stewardship of forests and to help people."

After retiring, Joseph continued to build the wealth he had acquired in forestlands and his stock portfolio, and he developed more concrete plans for philan-

Ibberson in his office at home, where many of his awards are displayed.

Dean of Agricultural Sciences Robert Steele presents a crystal Nittany Lion to Ibberson for establishing two endowed chairs in the School of Forest Resources.

thropy. But he didn't rest on his laurels. He continued to manage his tree farms and also engaged in forestry consulting, and had more time to relax while gardening.

Several friends and others had approached Ibberson about managing their forestlands, and they availed themselves of his expertise after he retired. Ibberson saw this as another opportunity for income that could be invested for later philanthropy. Within a short time, he was managing timber sales for five clients. His first client owned more than 1,000 acres, all of it ready for timber harvests. The second client owned 850 acres of forestland, most of which was ready for timber sales. He employed the consulting firm of Bommer and Geisman to assist with expert timber marking, and at times they assisted with timber sales contracts.

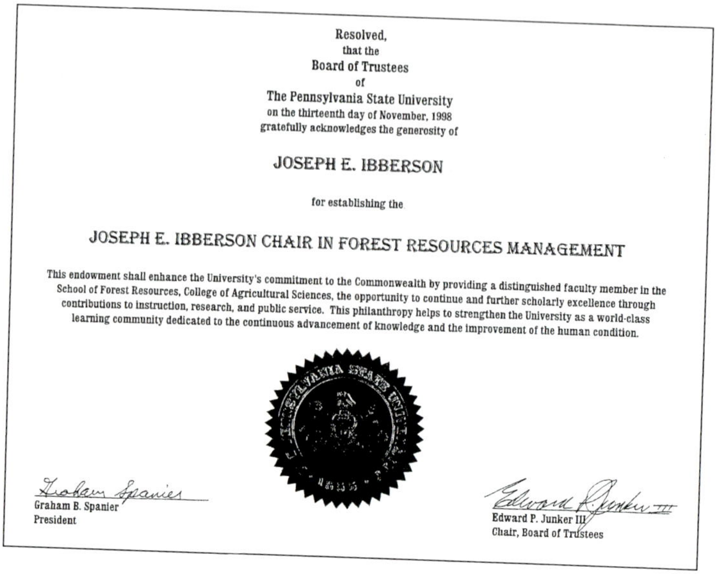

Resolved,
that the
Board of Trustees
of
The Pennsylvania State University
on the thirteenth day of November, 1998
gratefully acknowledges the generosity of

JOSEPH E. IBBERSON

for establishing the

JOSEPH E. IBBERSON CHAIR IN FOREST RESOURCES MANAGEMENT

This endowment shall enhance the University's commitment to the Commonwealth by providing a distinguished faculty member in the School of Forest Resources, College of Agricultural Sciences, the opportunity to continue and further scholarly excellence through contributions to instruction, research, and public service. This philanthropy helps to strengthen the University as a world-class learning community dedicated to the continuous advancement of knowledge and the improvement of the human condition.

Graham B. Spanier
President

Edward P. Junker III
Chair, Board of Trustees

Above: *Penn State acknowledged establishment of the Chair in Forest Resources Management.* Left: *School of Forest Resources Director Charles Strauss presenting the 2003 Outstanding Alumnus Award to Ibberson.*

Ibberson checked the timber-harvesting areas every week for compliance with contract specifications.

Both the consulting fees Ibberson earned and the proceeds from his own tree farms were substantial. Most of the timber sales for clients were valued at more than $100,000 each. So as Joe expressed it, "The funds for later philanthropy were really rolling in." He continued to buy forest properties whenever the opportunity arose and to manage these lands. Joseph gradually withdrew from the consulting business because other activities were taking more of his time, particularly managing his portfolio of common stocks, but he did continue to assist and advise a few close friends.

Nearly all of Joseph's investments in the stock market proved to be successful. The value of his stock portfolio was increasing dramatically, and dividends were accumulating. His retirement pension was more than he needed for living expenses. His pension had appreciable value because of his long service in the State Employees Retirement System while employed in the Bureau of Forestry, plus the years he spent in military service. Social Security payments further added to his income when he became eligible. So he was well off financially after he retired, and enjoyed good health too.

Public accolades were not forthcoming, however, until after Ibberson's retirement. Perhaps his unpretentious, quiet demeanor and laid-back management style

Mary Jane Davis, vice chairman of the Middle Paxton Board of Supervisors, honors Ibberson for donating the first ever township conservation Area.

had something to do with the delayed recognition. But then the magnitude of his accomplishments began to dawn on the people who knew him or heard about him, and many forms of recognition subsequently have been heaped on him.

Formal recognition of Ibberson's substantial accomplishments did not come until after he retired from the Bureau of Forestry. Then, as people became increasingly aware of his achievements in forestry and gifts to his profession, the public, and education, he received more and more awards and other forms of recognition.

The tributes acknowledged several kinds of achievements. Some awards were for his work in conserving Pennsylvania's forests, wildlife, and other natural resources. The excellence of his tree farms also was celebrated, as he has been a strong supporter of the Tree Farm Program. His donations of land were greatly appreciated, as were his gifts to Penn State that endowed professorship chairs. Ibberson's personal qualities were mentioned too, including generosity, innovation, keenness of mind, inspiration, and commitment to conserve natural resources and enrich lives through education.

Another form of recognition that is very meaningful to Joseph consists of unsolicited comments from family members and acquaintances. One of these whom Joe is especially proud of is a nephew, Glenn Ferguson, who is president and CEO of the Clark Construction Corporation. The company had current contracts in 2005 for $740 million to remove and replace 4,000 World War II dwellings at Fort Benning, Georgia; another for $650 million to remove and replace the high officers' quarters at Fort Belvoir, near Washington, D.C.; and yet another for a major project at the San Diego, California, naval base. At age forty-six, Ferguson heard indirectly about the Ibberson biography being written and jotted down the following thoughts (slightly edited), which give us some further insight into Joe's personality:

> "I have known Uncle Joe for as long as I can remember. My earliest memories are visiting him at a house he used to own that was located overlooking the Susquehanna River. As a kid, I thought it was one of the most amazing homes—it had unbelievable views for miles and miles of the river, there was a pond out back

stocked with fish, and lots of wildlife around. Uncle Joe would always take time to explain everything about the outdoors—different tree and plant types, the fish, the animals and their behavior—I always found it fascinating as a kid that someone could know so much about this. I remember thinking that trees were just trees, and why would someone bother with naming so many different kinds. But Uncle Joe would usually answer my endless questions, and I found my visits entertaining.

"During my teenage and college years I lost touch with Uncle Joe. I did not visit my grandmother [Joe's sister] so much during those years. After college, I would

Awards and Other Forms of Recognition

1977	**Wildlife Conservation Award, Pennsylvania Game Commission**
1984	**Outstanding Tree Farm of the Year in Area VI, American Forest Foundation**
1998	**Outstanding Tree Farmer of the Year, American Forest Foundation**
1998	**Acknowledgement for the Joseph E. Ibberson Chair in Forest Resources Management, Pennsylvania State University Trustees**
1999	**Mount Nittany Society Member, Pennsylvania State University**
1999	**Outstanding Management Resources Award, Northeastern Loggers Association**
1999	**Special Recognition Award for Forest Stewardship Conservation, Dauphin County Conservation District**
2000	**Commendation for Major Accomplishments in the Preservation and Conservation of the Commonwealth's Natural Resources, Senate of Pennsylvania**
2000	**Special Recognition Award for Conservation Stewardship, Dauphin County Conservation District**
2000	**Laurel Circle Award, Pennsylvania State University**
2000	**Proclamation of congratulations, Senate of Pennsylvania**
2001	**Elected Fellow, Society of American Foresters**
2001	**Dr. Joseph T. Rothrock Conservationist of the Year Award, Pennsylvania Forestry Association**
2001	**Pennsylvania Landowner of the Year Award, Nature Conservancy, Audubon Society, and Pennsylvania Wildlife Federation**

visit my grandmother once or twice a year, and during those visits, I would usually see Uncle Joe. So I guess I got reacquainted with Uncle Joe during those years. I got to learn that he was more than someone who just knew a lot about the woods!

"I also learned more about Uncle Joe's experiences in the Army, his life growing up, the Depression, some interesting facts about my grandmother and Uncle Joe going to college. Uncle Joe did not really open up much at first, but over the years, he has opened up more and more. When I read the book, *The Greatest Generation* or read various articles about his generation, Uncle Joe certainly fits many

2001	**Citation for major accomplishments, Pennsylvania House of Representatives**
2002	**Award in appreciation for donating a conservation area to the citizens of Pennsylvania, Department of Conservation and Natural Resources**
2003	**Outstanding Alumni Award, School of Forest Resources, Pennsylvania State University**
2003	**Recognition for extraordinary generosity and inspirational commitment to enrich lives through education and cultural preservation, Pennsylvania State University**
2003	**Award at a luncheon in the home of President Graham Spanier, in appreciation for the Joseph E. Ibberson Chair in Forest Resources Management, College of Agricultural Sciences, Pennsylvania State University**
2004	**Centennial Fellows Award, Penn State Mont Alto**
2004	**Award in grateful appreciation for donating the Stoney Creek land as a township conservation area**
2004	**Conservationist of the Year Award, Central Pennsylvania Conservancy**
2005	**Certificate of tribute from Harrisburg Mayor Stephen R. Reed for outstanding legacy of service and contributions to Pennsylvania**
2006	**Recognition honoring Joseph E. Ibberson's distinguished service for pioneering commitment to the forests of Pennsylvania and the nation and for his visionary leadership for future generations of foresters, Pennsylvania State University**
2006	**Outstanding Alumni Award of the College of Agricultural Sciences**

of the general descriptions. I thought he could have been in *The Greatest Generation* as I was reading the book. As he (and I) got older, he shared more experiences. He would actually start to bring photographs along or show some old photographs. Each visit, he would seem to cover a different topic.

"I remember one time he brought along some old photographs of the construction of a dam, and I learned that Uncle Joe had worked on the project. I had never

Roy Siefert, president of the Pennsylvania Forestry Association, congratulating Ibberson and presenting the Joseph T. Rothrock Conservationist of the Year Award.

Ibberson received the 1998 Outstanding Tree Farmer of the Year Award from the American Forest Foundation, flanked by John Dougherty (left) and Senator Roger Madigan.

known that he worked construction until then. (Since I am in construction, this was of particular interest.) I was fascinated by the construction equipment and techniques, which Uncle Joe explained in detail. Labor was obviously much cheaper back then, and I could not believe the amount of labor on the job versus the amount of equipment, as compared to today.

"It was also during this time that I started having children in my life. I was surprised by how good Uncle Joe was with kids, since he did not have any of his own. He was always gentle with the kids and has a lot of patience, and they were all warm to him. One of my favorite stories about Uncle Joe was when we went fishing with my son Ryan, who was four years old at the time. We went fishing at the second pond at the tree farm, where there was a picnic table. Uncle Joe would cast the rod for Ryan, then give the pole to Ryan to hold while we sat at the table. All of a sudden Ryan catches a big one. Uncle Joe helps Ryan start to reel in the fish, and then the bass jumps above the water. Well, Ryan is terrified—he thinks that the bass is going to get him—so he drops the fishing rod and starts running. I have never seen Uncle Joe laugh so hard. We convince Ryan that the fish will not get him, so he comes back and holds the rod again. Then Uncle Joe and Ryan start to reel the fish in again—and sure enough, it jumps out of the water and Ryan again drops the rod and runs. This went on and on for a while, before we finally got the fish to shore. I actually got most of this on videotape, so it is no fish story. Whenever we see Uncle Joe, we laugh about that every time.

"I also enjoy hearing Uncle Joe's stories about when he was a kid and growing up, when they did not have TVs, radios, microwaves, or computers. I make sure my kids listen to that, since they find it unbelievable that you could grow up without all of those things. I really want my kids to hear about it firsthand, since they do not believe me when I tell them. I think this is something that is nice to pass on to future generations.

"I also learned that Uncle Joe was a shrewd businessman, and a good investor in real estate, which is obviously how he could afford to purchase so much land. He has the classic investment philosophy that many do not follow in today's world of fast-paced trading. I recall Uncle Joe telling me he has owned one stock for fifty years, and the yearly dividend is now more than he paid for the stock. I have learned from listening to him to stick to the basics—buy a good stock, then hold on to it. For a person that does not use the Internet, it is amazing how much Uncle Joe knows about the stock market. I also invest in real estate and the stock market, and I have learned a lot from listening to Uncle Joe, especially about buying and holding.

"Uncle Joe lost his first wife many years ago, and he was really fortunate to develop a relationship with Nila. They had been neighbors for almost forty years, I think, and then they both lost their spouses. Nila and Uncle Joe seem to be really compatible, and I know that whenever I have to track down Uncle Joe, it is easier to reach him at Nila's house than at his house. He does not have an answering machine.

Senate of Pennsylvania proclamation congratulating Ibberson.

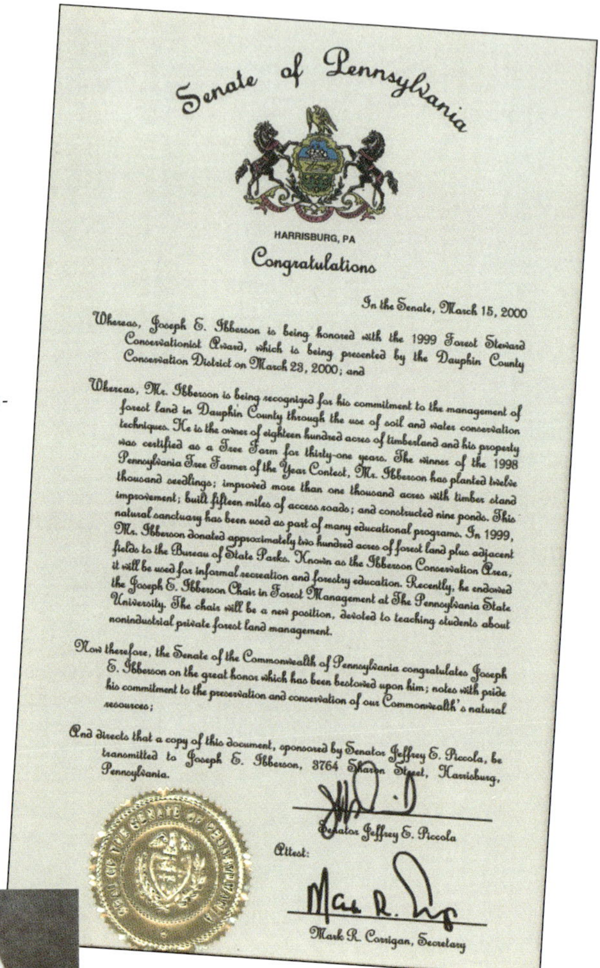

Senate of Pennsylvania

HARRISBURG, PA

Congratulations

In the Senate, March 15, 2000

Whereas, Joseph E. Ibberson is being honored with the 1999 Forest Steward Conservationist Award, which is being presented by the Dauphin County Conservation District on March 23, 2000; and

Whereas, Mr. Ibberson is being recognized for his commitment to the management of forest land in Dauphin County through the use of soil and water conservation techniques. He is the owner of eighteen hundred acres of timberland and his property was certified as a Tree Farm for thirty-one years. The winner of the 1998 Pennsylvania Tree Farmer of the Year Contest, Mr. Ibberson has planted twelve thousand seedlings; improved more than one thousand acres with timber stand improvement; built fifteen miles of access roads; and constructed nine ponds. This natural sanctuary has been used as part of many educational programs. In 1999, Mr. Ibberson donated approximately two hundred acres of forest land plus adjacent fields to the Bureau of State Parks. Known as the Ibberson Conservation Area, it will be used for informal recreation and forestry education. Recently, he endowed the Joseph E. Ibberson Chair in Forest Management at The Pennsylvania State University. The chair will be a new position, devoted to teaching students about nonindustrial private forest land management.

Now therefore, the Senate of the Commonwealth of Pennsylvania congratulates Joseph E. Ibberson on the great honor which has been bestowed upon him; notes with pride his commitment to the preservation and conservation of our Commonwealth's natural resources;

And directs that a copy of this document, sponsored by Senator Jeffrey E. Piccola, be transmitted to Joseph E. Ibberson, 3764 Sharon Street, Harrisburg, Pennsylvania.

Senator Jeffrey E. Piccola

Attest:

Mark R. Corrigan, Secretary

Governor Tom Ridge congratulates Ibberson for the Pennsylvania Landowner of the Year Award 2001.

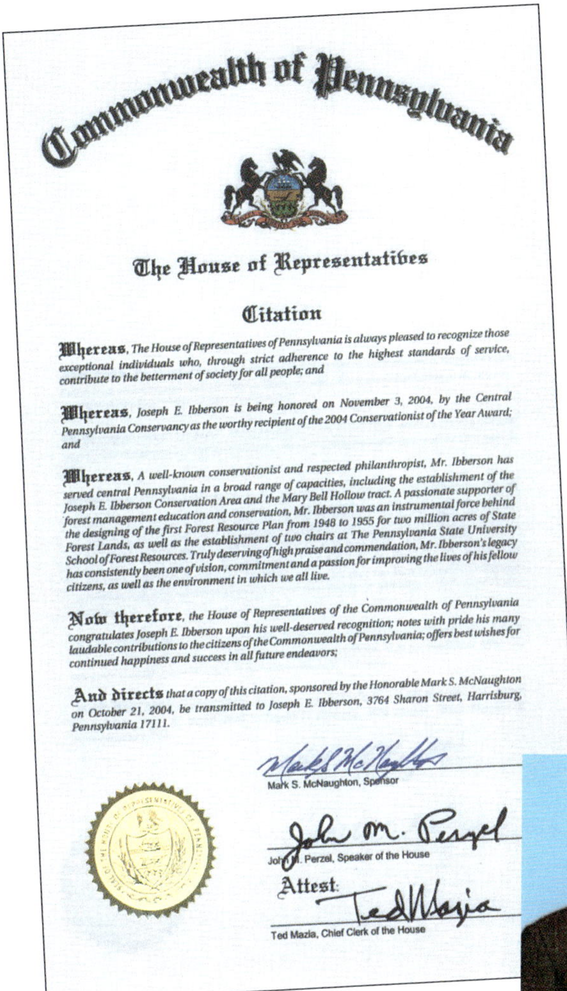

Commonwealth of Pennsylvania

The House of Representatives

Citation

Whereas, The House of Representatives of Pennsylvania is always pleased to recognize those exceptional individuals who, through strict adherence to the highest standards of service, contribute to the betterment of society for all people; and

Whereas, Joseph E. Ibberson is being honored on November 3, 2004, by the Central Pennsylvania Conservancy as the worthy recipient of the 2004 Conservationist of the Year Award; and

Whereas, A well-known conservationist and respected philanthropist, Mr. Ibberson has served central Pennsylvania in a broad range of capacities, including the establishment of the Joseph E. Ibberson Conservation Area and the Mary Bell Hollow tract. A passionate supporter of forest management education and conservation, Mr. Ibberson was an instrumental force behind the designing of the first Forest Resource Plan from 1948 to 1955 for two million acres of State Forest Lands, as well as the establishment of two chairs at The Pennsylvania State University School of Forest Resources. Truly deserving of high praise and commendation, Mr. Ibberson's legacy has consistently been one of vision, commitment and a passion for improving the lives of his fellow citizens, as well as the environment in which we all live.

Now therefore, the House of Representatives of the Commonwealth of Pennsylvania congratulates Joseph E. Ibberson upon his well-deserved recognition; notes with pride his many laudable contributions to the citizens of the Commonwealth of Pennsylvania; offers best wishes for continued happiness and success in all future endeavors;

And directs that a copy of this citation, sponsored by the Honorable Mark S. McNaughton on October 21, 2004, be transmitted to Joseph E. Ibberson, 3764 Sharon Street, Harrisburg, Pennsylvania 17111.

Mark S. McNaughton, Sponsor

John M. Perzel, Speaker of the House

Attest:

Ted Mazia, Chief Clerk of the House

House of Representatives citation congratulating Ibberson.

Governor Ed Rendell thanks Ibberson in 2005 for all of his good deeds for Pennsylvania.

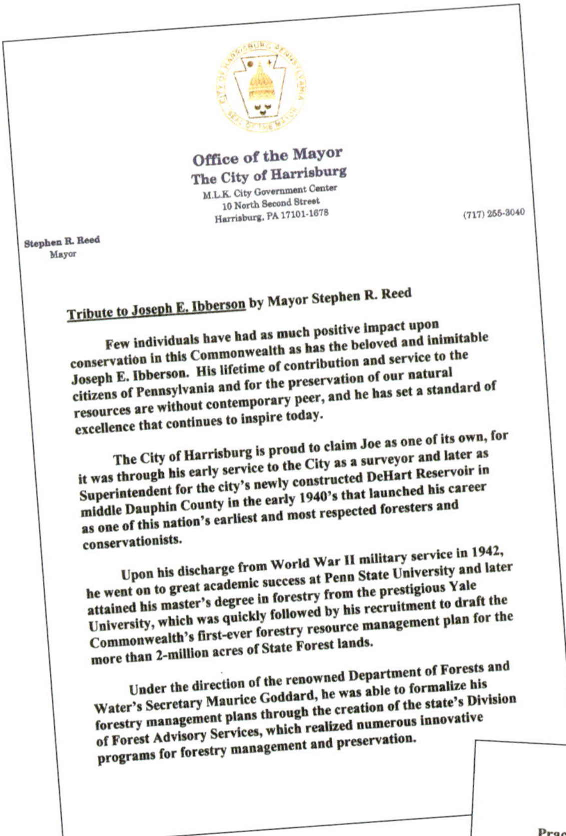
Office of the Mayor
The City of Harrisburg
M.L.K. City Government Center
10 North Second Street
Harrisburg, PA 17101-1678
(717) 255-3040

Stephen R. Reed
Mayor

Tribute to Joseph E. Ibberson by Mayor Stephen R. Reed

Few individuals have had as much positive impact upon conservation in this Commonwealth as has the beloved and inimitable Joseph E. Ibberson. His lifetime of contribution and service to the citizens of Pennsylvania and for the preservation of our natural resources are without contemporary peer, and he has set a standard of excellence that continues to inspire today.

The City of Harrisburg is proud to claim Joe as one of its own, for it was through his early service to the City as a surveyor and later as Superintendent for the city's newly constructed DeHart Reservoir in middle Dauphin County in the early 1940's that launched his career as one of this nation's earliest and most respected foresters and conservationists.

Upon his discharge from World War II military service in 1942, he went on to great academic success at Penn State University and later attained his master's degree in forestry from the prestigious Yale University, which was quickly followed by his recruitment to draft the Commonwealth's first-ever forestry resource management plan for the more than 2-million acres of State Forest lands.

Under the direction of the renowned Department of Forests and Water's Secretary Maurice Goddard, he was able to formalize his forestry management plans through the creation of the state's Division of Forest Advisory Services, which realized numerous innovative programs for forestry management and preservation.

Harrisburg Mayor Stephen Reed's tribute to Ibberson.

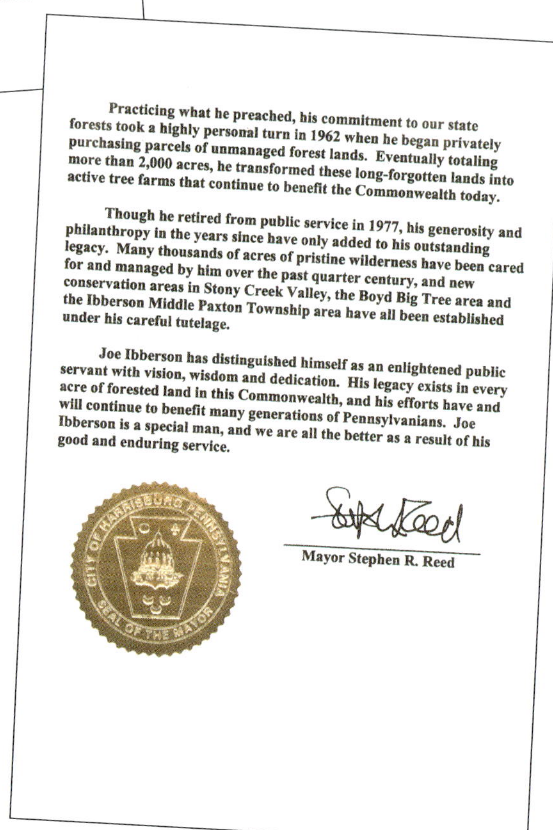
Practicing what he preached, his commitment to our state forests took a highly personal turn in 1962 when he began privately purchasing parcels of unmanaged forest lands. Eventually totaling more than 2,000 acres, he transformed these long-forgotten lands into active tree farms that continue to benefit the Commonwealth today.

Though he retired from public service in 1977, his generosity and philanthropy in the years since have only added to his outstanding legacy. Many thousands of acres of pristine wilderness have been cared for and managed by him over the past quarter century, and new conservation areas in Stony Creek Valley, the Boyd Big Tree area and the Ibberson Middle Paxton Township area have all been established under his careful tutelage.

Joe Ibberson has distinguished himself as an enlightened public servant with vision, wisdom and dedication. His legacy exists in every acre of forested land in this Commonwealth, and his efforts have and will continue to benefit many generations of Pennsylvanians. Joe Ibberson is a special man, and we are all the better as a result of his good and enduring service.

Mayor Stephen R. Reed

"Uncle Joe is also the busiest retired person that I know. I cannot believe how hard he works at this point in his life, and I know that he takes tremendous pride in his tree farm. I have enjoyed watching Uncle Joe explain the tree farm to my kids, and I see how much pleasure he takes in it. I think he really wants to pass a lot of his knowledge on to future generations. I also like to watch some of his 'old tricks' that he tries on each kid. He used to tell Ryan, who is now 11 and does not fall for it anymore, that there were bears in a rocky part of the tree farm. We would then walk around looking for them, and Uncle Joe would always make some kind of noise to scare Ryan and send him running.

"A few years ago, my grandmother passed away, and Uncle Joe was helpful in dealing with some difficult family issues. It also means that he is the last of his siblings still living. We have not seen as much of him since my grandmother passed away, but every time we do, we make sure he tells us some of his stories and experiences so that they are not lost."

Glenn Ferguson's mother, Beverly Stutzman Ferguson, the daughter of Joe's sister, Ruth, also sent a note with some comments about Joe.

"Much has been written, awards received, and said about Joseph Ibberson's professional life, but what about his private life? Uncle Joe is a generous, kind, thoughtful, and understanding individual. When asked, he will listen to your problem and give valuable advice.

"Some things in life are inherited, like the love of gardening. Uncle Joe's mother always had a vegetable and flower garden every spring. Although I think he no longer plants carrots, when I was a child he would pull the carrots, wash them off, and I would eat them. We decided that is why I have good eyesight.

" During World War II, Uncle Joe always remembered me with gifts—a beautiful ring and a pair of gloves that I had saved and my daughters also wore.

"After finishing a forestry course at Penn State, Uncle Joe helped build and became caretaker of the Clark's Valley Dam. I have happy memories of our numerous visits to the dam.

"The tree farm is a magnificent tribute to Uncle Joe's dream and vision. We enjoy visiting there because it is a place of beauty, serenity, and peace.

"His biggest legacy is the state park that bears his name. Generations of children and adults will enjoy, learn, and benefit from his generosity and unwavering faith in preserving the land and making the world a better place."

Some of the key people who worked with Ibberson in developing his tree farms also wrote about their relationships with him. The Petroski family have known him for thirty years, and became close friends and confidants (Appendix D). Ben Mentzer has known Joe nearly as long, and they too became good friends (Appendix E). Their reminiscences amplify an understanding of Joe's personality and character.

Above: *Joe's garden, where he grows only the most healthful varieties of vegetables and fruits, viewed from the patio behind his home.* Left: *As Joe explained why every garden should have a wren house to control bugs, a wren flew in with worms in its beak.*

In retirement, Joe was able to devote more time to gardening, one of his favorite activities. He had learned about home gardening from his mother starting at an early age, perhaps when he was less than ten years old. She was born and raised on a farm, so with that background she became an excellent gardener and passed her knowledge on to her attentive son. Ibberson has had his own home garden since moving to Harrisburg. He grows the most healthful varieties of fruits and vegetables, using some unusual gardening techniques.

Joe's garden produces amazing quantities of delicious, colorful, and nutritious vegetables and fruits throughout the growing season. Asparagus spears pop out of the ground in early spring. Other abundant crops of vegetables packed with healthy nutrients continue through the summer, until the last of the kale is dug from the snow in December. Some of the more unusual vegetable varieties he grows include red onions, ruby-colored corn, Health Kick tomatoes with 50 percent more lycopene than other varieties, purple carrots, Yukon gold and blue potatoes, purple pole beans, and dark purple, orange, and golden peppers. Joe makes salads from the vegetables daily from the middle of June to November.

Fruits also are harvested and consumed over many months. By the middle of spring, there is always a bumper crop of delicious red strawberries. Ever-bearing gold raspberries follow, continuing into November. Flavorful apricots ripen in June, followed by Shiro yellow plums and then Japanese red plums. Joe stores apples in the refrigerator so that a good supply remains available through January.

The colorful varieties not only are attractive, but the flavonoids and other pigments they contain make the fruits and vegetables extra healthful. There is medical evidence that the green, yellow, orange, red, blue, and purple pigments fight heart disease, cancer, inflammation, obesity, elevated blood sugar, and memory loss.

Ibberson grows all of the vegetables and fruits organically, not using pesticides on any of them. To help keep the garden free of insects throughout the summer, Joe erected and maintains two wren houses. The wrens return here to nest every year. To feed

Right: *At age eighty-eight, Joe was still planting young fruit trees and anticipating future harvests.* Below: *Garden paths made with twelve-inch blocks, for easy planting, maintenance, and harvesting.*

Above: *Joe finds it relaxing to watch from his patio as thousands of butterflies, bees, and wasps flit about his home-grown butterfly weeds.* Left: *Ever-bearing golden raspberries are harvested throughout the growing season.*

their babies, the mother wrens scour the garden to find all the insects they require. In some years, after the wrens depart for warmer climates, cabbage worms may appear on the broccoli and kale. Every year Joe spreads decomposed compost over the garden from a compost pile at the rear. Occasionally he adds Canadian peat to the garden where needed. He maintains a circle of white clover under each fruit tree, which supplies nitrogen and reduces maintenance.

While gardening, Ibberson often gets visitors from the neighborhood, and fields many questions from them about his hobby. And somehow they always manage to leave with a bag of fresh produce, with admonitions from Joe about healthful eating (see Appendix F). Ibberson thinks every home should have a garden. At the age of eighty-eight he was still planting young fruit trees, growing and harvesting the bounty of his garden, and sharing his produce with others to help keep them and himself healthy.

CHAPTER 19

Nila

About a year after Evon's death, another lady became prominent in Joe's life. Nila L. Grimes, a longtime neighbor, had lost her husband in 1990. Like Joe, Nila too had been caring for her spouse during protracted years of illness, so it was a blessing that they could comfort each other. They quickly realized they had much in common.

Nila and Earl F. Grimes had moved into a house across the street from the Ibberson residence in 1951, a few months before Evon and Joe occupied their newly built house. They soon became acquainted as neighbors, but beyond that the two families did not socialize, as all four of them were busy with their careers and their own circles of acquaintances. Joe and Nila did not foresee that they would become much closer friends four decades later.

Nila spent her early years in Star Junction, a small rural town in southwestern Pennsylvania between Pittsburgh and Uniontown. She often played outdoors and loved nature. At the age of eight, Nila moved to Harrisburg with her parents and sister. After finishing high school, Nila found employment in the credit office of Bowman's Department Store. Later she worked for twenty-seven years as an office manager at the New Cumberland Army Depot.

Nila's father, Arthur Sillett, had come to America in about 1891 from Durham, England. Durham is about thirty miles north of Huddersfield, where Joe's grandfather William Ibberson had lived before emigrating to America.

Nila really didn't get to know Joe well until 1992. They had exchanged pleasantries over the years when they chanced to meet in the neighborhood or in a grocery store but really had not become well acquainted. They hit it off right away, discovering they both had grown up in coal country and had similar interests.

Their initial friendship quickly developed into a dedicated partnership. He came to think of her as not only a congenial neighbor, but a very attractive, intelligent lady who shared his outlook on life, nature, gardening, and politics.

For forty years, on most holidays and weekends, Nila had observed Joe loading tools and heading out in his car. He took along a surveying transit, chain saws, pruners, shovels, and other tools for long days in the forests. During hunting seasons, Joe carried a rifle or shotgun. Nila assumed that Joe merely had a hunting cabin and a few acres of forestland. It was not until 1992 that Joe invited Nila to visit his tree farms.

Nila was astonished to see about 2,000 acres that Joe had conserved and managed. She marveled at the timber stand improvement projects accomplished on hundreds of acres. As they drove over some twenty-five miles of forest roads on his properties, they paused next to several ponds to observe fish, birds, and various animals. They looked out

Left: *Nila communing with her favorite tree on the Alpha Tree Farm.* Below: *Nila was enthralled by the fabulous vista from the Alpha Tree Farm—the Susquehanna River, valleys, and ridges, thirty-six miles to the horizon.*

Nila shared the limelight with Joe when he received the 2001 Landowner of the Year Award.

from a fabulous vista across miles of breathtaking landscapes stretching to the Susquehanna River and passed numerous picnic areas before stopping at a rustic table for lunch. Nila's astonishment grew into admiration for Joe's accomplishments and led to a new, revitalizing chapter in Joe's life.

Loving the outdoors as she does, Nila has enjoyed frequent trips with Joe to the Alpha Tree Farm. She especially likes taking wildlife safaris over the twelve miles of shaded roads. Nila is particularly impressed with the large trees, especially one large, tall, straight tulip poplar that is her favorite. She calls it "the gentleman of the forest" and never fails to pay it a visit during trips to the tree farm.

Nila has expressed concern over what will happen to the Alpha Tree Farm after Ibberson no longer will be able to manage and care for it. If she had her way, the innocent, beautiful wildlife would be enduringly protected and not hunted.

Both Joe and Nila have been devout students of the Bible. As their relationship became closer, Nila started giving booklets to Joe that suggested daily readings from the Bible by chapter and verses. He questioned the value of this at first, as he already was in the habit of reading the Bible occasionally. Nila persevered, however, and would check to see if Joe had completed the daily reading. Sometimes then they would discuss its meaning and ask themselves if their lives were following the scriptures. Nila is a true Christian. That is part of her inner beauty, and it beams from her face. She is devoted to helping others, even when it creates hardship for herself.

Ibberson became so familiar with the Bible, after looking up chapters and verses for many years, that he no longer had to use the table of contents or the Tyndale Bible Verse Finder. Sometimes he would open the Bible to the exact passage for that day. Ben Mentzer, who works at the Alpha Tree Farm, is another student of the Bible. He seems to know it from beginning to end. Formerly when the subject came up, Ibberson used to do all the listening. But after Nila's influence on Joe's reading habits, he then engaged Ben more readily in a two-way conversation about biblical subjects.

As part of their partnership activities, Nila has been assisting Joe with administrative matters. She checks to see that he answers the mail, pays the taxes, attends meetings and meets guests on time, and takes phone messages. She even makes sure that he orders the right size of stone for road construction projects at the tree farms. Ibberson does not have an answering machine and thinks most callers would rather talk with Nila anyway.

Nila is a true blue-and-white Penn Stater, and she has been a loyal football fan. She attended all the games while her nephew Roger Grimes was the varsity starting tailback for the Nittany Lions during 1965 to 1967, and she continued to come to home games sporadically afterward. Joe also has maintained his interest in Penn State football. In 2003, he responded to a dinner invitation from Joe and Sue Paterno who had common interests in philanthropy to the university as well as football.

Her favorite television programs have mostly been sitcoms, such as *Everybody Loves Raymond, The Andy Griffith Show, All in the Family,* and *The Honeymooners.* She also likes stand-up comedians such as Jack Benny, Carol Burnett, Red

Carol Strauss, wife of School of Forest Resources Director Charles Strauss, smiles with Joe and Nila in appreciation of their generosity.

Nila Grimes, flanked by Joe Ibberson and Henry Gerhold, provided biographical information through interviews.

Skelton, Bob Hope, and Flip Wilson. Any of these would preempt the Super Bowl or Nila's favorite John Wayne movies, *The Quiet Man* and *Rio Bravo.* Nila also enjoys educational programs, especially if they relate to the environment or religion.

Nila has a great sense of humor, and she and Joe often tease each other. She has many friends, and according to Joe, she receives enough greeting cards to keep a card shop in business.

Knowing that she would be interviewed, she commented in 2004, "I like living here, I like the neighbors." With a wink, she let it be known that by "neighbors," she really was referring to Joe. Some months later, Nila jotted down some thoughts about Joe: "I admire his quick wit and warm personality, his caring and understanding ways. I respect his keen, calm insight and tough-minded approach to any problem. Before making a decision he always has the facts to back up his approach. And then away he goes. He has character; that is, doing what is right when no one is looking. For him, life is good, and he enjoys each day."

Ibberson stated that he thanks the Lord every day for such a great lady as Nila for him to love and to be a full partner in his activities and philanthropic endeavors.

CHAPTER 20

Life's Goals Fulfilled

It is fitting to recall that it was the story of Joseph Rothrock that inspired Joe Ibberson to pursue a career in forestry. Gifford Pinchot paid tribute in Rothrock's eulogy, in which he recounted the man's many achievements in forestry, medicine, and other kinds of public service. Here is the first part of Pinchot's remarks:

> Dr. Joseph Trimble Rothrock was first of all a great public servant. Unselfish to the point of extreme self-sacrifice, capable to the level of the brilliant achievements which distinguished his career, the thread upon which his innumerable activities were strung was that of selfless service to the common welfare.
>
> Dr. Rothrock was always a pioneer. He was one of the few men of his generation whose thoughts customarily went before that of his contemporaries, and laid the ground for great advance in the commonwealth he loved so well. At every turn in the story of his uniquely useful life, we find things done that no one has done before.

These initial words of Pinchot could apply also to Joseph E. Ibberson, though the particulars of their careers differ. Ibberson and Rothrock both were great public servants devoted to Pennsylvania's forests, both had pioneering ideas and great achievements, and both were motivated by selfless service to the common welfare.

Joseph Ibberson's accomplishments have not been as well known as Rothrock's until recently. But more and more, people will come to appreciate all that Ibberson has done for the professional management of Pennsylvania's forests, for all to enjoy and for forestry education. And they will realize that he knew how to live a productive, healthy, and satisfying life, which he has shared selflessly with others.

Ibberson's Favorite Quotes

Following are some of Joseph Ibberson's favorite quotes from scriptures and well-known people. All scriptures are taken from the New Living Translation. The quotes influenced Ibberson's life and aspirations.

- **There are three things that will endure—faith, hope, and love—and the greatest of these is love.** *I Corinthians 13:13*
- **Do for others what you would like them to do for you. This is a summary of all that is taught in the law and the prophets.** *Matthew 7:12*
- **Love does no wrong to anyone, so love satisfies all of God's requirements.** *Romans 13:10*
- **Intelligent people are always open to new ideas. In fact, they look for them.** *Proverbs 18:15*
- **Zeal without knowledge is not good; a person who moves too quickly may go the wrong way.** *Proverbs 19:2*
- **Just as the body is dead without a spirit, so also faith is dead without good deeds.** *James 2:26*
- **If you can dream it you can do it.** *Walt Disney*
- **Conservation is the application of common sense to the common problems for the common good.** *Gifford Pinchot*
- **The vast possibilities of our great future will become realities only if we make ourselves responsible for that future.** *Gifford Pinchot*
- **The nation behaves well if it treats the natural resources as assets which it must turn over to the next generation increased, and not impaired in value.** *Theodore Roosevelt*
- **The richest Conservation is a state of harmony between men and land.** *Aldo Leopold*
- **We abuse land because we regard it as a commodity belonging to us. When we see land as a community to which we belong, we may begin to use it with love and respect.** *Aldo Leopold*
- **The clearest way into the universe is through a forest wilderness.** *John Muir*

- **Society speaks and all men listen, mountains speak and wise men listen.** *John Muir*
- **The mountains are calling and I must go.** *John Muir*
- **How glorious a greeting the sun gives the mountains!** *John Muir*
- **Our lives . . . need the relief of where the pine flourishes and the jay screams.** *Henry David Thoreau*
- **A town is saved, not more by the righteous men in it than by the woods and swamps that surround it.** *Henry David Thoreau*
- **The greatest wonder is that we can see these trees and not wonder more.** *Henry David Thoreau*
- **God bless America. Let's save some of it.** *Edward Abbey*
- **Those who contemplate the beauty of the earth find reserves of strength that will endure as long as life lasts.** *Rachel Carson*
- **The only way to keep your health is to eat what you don't want, drink what you don't like, and do what you'd rather not.** *Mark Twain*
- **Even if you're on the right track, you'll get run over if you just sit there.** *Will Rogers*
- **No man is great if he thinks he is.** *Will Rogers*
- **People are marvelous in their generosity if they just know the cause is there.** *Will Rogers*

Ibberson's friend and colleague Norman Lacasse wrote in 2003:

> Mr. Ibberson's generosity is legendary. His gift to us, a long career of dedicated and innovative service, will not only benefit us but our children and our children's children. He stands as a giant redwood in the forest. Thank you, Joe!

In April 2006, Penn State President Graham Spanier and Joe Paterno presented a fitting tribute that summarizes Ibberson's contributions and legacy of his career and philanthropy:

As President of the Pennsylvania State University, I am honored to present
JOSEPH E. IBBERSON
With this recognition of
DISTINGUISHED SERVICE
For His Pioneering Commitment to the Forests of Pennsylvania and The Nation
and his Visionary Leadership for Future Generations of Foresters.

> Throughout the last six decades, Joseph E. Ibberson has transformed the profession of forestry through his tireless efforts to improve management and preservation practices throughout the Commonwealth and the United States. In an act of unparalleled generosity, he donated his own property to create the first conservation area in the Pennsylvania State Park System. At The Pennsylvania State University, he has established faculty chairs in both forest management and community and urban forestry, ensuring that students will receive the education and training necessary to uphold the high standard of excellence that Mr. Ibberson has set for the entire profession of forestry.

President Spanier's tribute epitomizes the legacy of forester Ibberson's life for the forestry profession, forest landowners, and nature lovers of all kinds.

Coach Joe Paterno presenting a Distinguished Service recognition to Joe Ibberson.

1955	Started volume table project
	Placed in charge of state forest tree nurseries
1956	Research Division changed to Forest Advisory Services
	Organized hiring and training of service foresters
1957	Hired wood utilization advisor
1962	Hired tree improvement specialist
	Started land purchases for his tree farms
1973	Formed Division of Forest Pest Management
1977	Retired from Bureau of Forestry
1991	Evanka Ibberson died
1992	Close relationship began with Nila Grimes
1996	Endowed gift to Pennsylvania Forestry Association
1998	Donated Ibberson Conservation Area to Department of Conservation and Natural Resources
	Endowed Forest Management Chair at Penn State, $1.5 million
2003	Committed $2 million to endow an Urban and Community Forestry Chair at Penn State
2004	Gave Ibberson Middle Paxton Conservation Area to Middle Paxton Township

APPENDIX B

Mexico Vacations

Joe and Evon spent many long vacations in Mexico during 1955 to 1978. Joe had been fascinated with the mysterious Mayas and other ancient tribes of Central America ever since he had studied history at Lykens High School. Where had they come from? Why did their advanced civilizations disappear? Will their hieroglyphics ever be deciphered and fully understood? At the time of the Ibbersons' travels, anthropologists were beginning to unearth some answers from their excavations under way in several cities and villages.

The knowledge of astronomy possessed by these ancients was the most advanced in the world at their time; even by today's standards, it was accurate and precise. Their agricultural practices also were amazing, even including plant breeding. Their religious beliefs and ceremonies, form of government, social life, and relationships among tribes all were different from those anywhere else in the world.

The Cinco de Mayo battle took place here in 1862.

To satisfy their curiosity, Joe and Evon traveled to many places in Mexico and as far as Guatemala. Joe had taken a course in Spanish, a language already familiar to Evon, so they could converse with the local people. Most of the trips took them by car from Harrisburg to the various places they explored in Mexico. They drove more than 5,000 miles within Mexico on one of their trips. In one remote area, the local people referred to them as *prima vera*—the first Americans they had ever seen. They did not consider the tourists to be gringos, rather heroes.

The water and food could be unsafe, except in first-class hotels. They could not trust dairy products because the pasteurizing equipment often broke down yet the bottling continued. Therefore, most meals consisted of cooked food and fruit that could be peeled. Bananas, papayas, and mangoes were in good supply and delicious. Before crossing the border to Mexico, Joe would pack the trunk of the Pontiac with crackers, cereals, granola bars, raisins, dried apricots, prunes, and other dry, nourishing food. Thus the couple stayed healthy in out-of-the-way places where only the fruit was safe to eat.

Such precautions became routine after a while. They refilled five one-gallon jugs with safe drinking water available at the better hotels. They used this bottled water also for brushing teeth, as the safety of water from the faucet was not dependable. The precautions definitely were preferable to the consequences of being careless. On the return trip, the first restaurant in the United States would present a welcome change. The milk always tasted so good.

The trips to Mexico were all quite long. The first leg of one trip extended from Harrisburg to Laredo, Texas. After provisioning the trunk with food and pure

The Tree of Tule, believed to be the oldest living tree in the world.

Evon clowning with five other birds.

water, they drove to Monterrey, then to Mexico City in central Mexico, from which they proceeded southeast via the Tehuantepec Peninsula to the Yucatan Peninsula. That would take seven days. On the return trip, they might spend some time at Acapulco, Puerto Vallarta, and Mazatlan along Mexico's west coast. A different and interesting return trip could pass through El Paso, Texas. The entire trip would take about thirty-five to forty-five days.

In Mexico City, they marveled at the Central American art that adorned the Mexico University Library. To the southeast was a breathtaking view of the Smoking Mountain, Popocatepetl, an active volcano that still erupts periodically. At Puebla was a reminder that the outnumbered Mexican Army defeated invading French troops on May 5, 1862, known as Cinco de Mayo.

Near Mitla, they came upon a huge Mexican cypress in the Valley of Oaxaca below Monte Alban. Growing in a churchyard in the village of Santa Maria del Tule, the tree is said to be the oldest living thing in the world. The tree was about 150 feet tall and 150 feet in circumference at the base. Two estimates of its age published in 1906 were 4,000 and 6,000 years.

Each successive trip included more of Mexico, until finally they had explored most of the country. At the higher elevations, they admired beautiful conifer forests. In the mountains it was always windy, and the native Indians wore serapes the year round to keep warm. The vegetation was entirely different in the sweltering swamps near Tehuantepec. They visited and explored practically all of the important historical and archaeological sites.

Fiesta parades were common in the small villages. Just about all the villagers dressed in the most colorful costumes, and they reveled in dancing to the beat of the music. It seemed that every day was a fiesta day somewhere in Mexico. Every village celebrates its Saint's Day with parades, religious ceremonies, dancing, and even fireworks.

The Mexicans were taken with Evon's exotic appearance and sometimes invited the Ibbersons to their homes. Joe and Evon enjoyed getting close to the people in

Joe having a great day fishing in the Pacific Ocean, where he caught forty-nine big ones.

rural areas, where they had dirt floors and bamboo furniture. They had to watch the water and the food, however. At one restaurant, they ordered fried eggs, which they presumed to be safe and readily available, but the proprietor had to send out for them.

They often stayed overnight in very primitive areas and occasionally were invited to spend an evening at the home of a local family. Joe recalled that one such home was in southern Mexico near the Guatemalan border. The thatched hut had a dirt floor, and all the furniture was made from reeds. Many relatives of the family were invited to meet and talk with "the Americanas." They wanted to know all about the United States, and the discussions lasted until 2 A.M. They asked the Ibbersons whether they were Catholic or Baptist, apparently believing that only Catholics were Christian. Joe and Evon tried to explain that they were Christians, but neither Baptist nor Catholic. The head of the family had one ambition, to send his oldest son to the University of Mexico to become a doctor. His brother asked to be taken along back to the United States and flashed a roll of American money to show that he could pay his own way. The family had a ring-tailed cat for a pet, which they passed around the entire evening.

On one trip, while traveling west on the Tehuantepec Peninsula, Joe and Evon bravely decided to take a detour on a narrow, rutted road through a very large swamp. As far as the eye could see was water interspersed with reeds and shrubs, apparently occupied only by wildlife. As they stopped to take a few photos, suddenly four almost naked Indians appeared nearby and advanced toward the automobile. Ibberson drove away hurriedly and stopped no more while traversing the swamp.

A contrasting event occurred while the Ibbersons were looking for overnight accommodations at Durango. They discovered that one of the homes of Dolores Del Rio, the movie actress, was available as a guest house, but before being accepted as guests, they had to be interviewed and approved. The interview was successful, the home was beautiful, and the dinner was one of the best.

On another occasion, they visited a compound of the notorious rebel leader Pancho Villa. The compound was well fortified and had rooms for many bodyguards. On display there was the bullet-hole-riddled Ford in which Pancho Villa had been assassinated on July 20, 1923. Later, while the body was being moved to a new grave, the head disappeared. One belief is that the head was entombed

Evon with Pancho Villa's wife, who took a shine to Evon.

at Yale University by a secret society called Skull and Bones. A Mexican writer asked President George H. W. Bush, a purported member of Skull and Bones, to return the head, but to no avail.

While the Ibbersons were at the compound, a young lady who was serving as a guide sometimes glanced furtively toward a door that was slightly ajar. Watching from behind the door was Luz Corral de Villa, the best known of Pancho Villa's many wives. (Reports indicate that he had as many as twenty-two wives and mistresses.) Luz Corral was awarded the property by the Mexican government after Pancho's death. On rare occasions, she would speak with visitors at her discretion, and she agreed to speak with the Ibbersons and to be photographed with them.

One of many mementos that Ibberson acquired was a large brass plate emblazoned with an Aztec calendar. It was made especially for him in Taxco, the oldest mining town in North America, by a silversmith he had befriended. Hernándo Cortés had discovered rich silver deposits there not long after his conquest of Mexico in 1521. In 1743, Joseph de la Borda, a French prospector, hit upon a rich vein of silver and amassed an immense fortune from Taxco silver mines. He built a large church and on the front inscribed, in letters five feet high, "THE LORD GIVES TO BORDA, AND BORDA GIVES TO THE LORD."

The Basilica of Our Lady of Guadalupa so impressed the Ibbersons that they visited it several times. Located a few miles north of Mexico City, this holy Roman Catholic shrine honors the Guadalupe Virgin, Mexico's patron saint. It is said that the Virgin in 1531 appeared on the site to a peasant Indian, Juan Diego, and directed him to build a church. The local bishop requested

Large Aztec calendar made especially for Joe by a silversmith he befriended at Taxco.

The Basilica of Our Lady of Guadalupa, one of Roman Catholicism's holiest shrines, where Joe saw many pilgrims with bloody knees from their mile-long penitence walk.

proof, so the Indian boy filled his cape with roses at the direction of the Virgin. When the bishop opened the cape, the roses disappeared and were replaced by an image of the Virgin on the cloth. The Ibbersons saw the cloth in a gold frame hanging above the altar. Pilgrims have contributed more than fifty tons of silver, which has been used for sacred vessels, altar rails, candelabra, and other accessories.

They also observed thousands of pilgrims hoping to have their sins forgiven perform a ritual penitence walk on their knees along a mile-long stone walkway leading to the Basilica. The seriousness of their sins determines how far from the Basilica they must start. Many endure bloody, injured knees by the time they reach the Basilica.

The Ibbersons visited and explored the major Mayan ruins, Teotihuacan, the City of Gods, about thirty-five miles north of Mexico City, as well as the Aztec civilization in the vicinity of the city. Joe's primary purpose on the many trips there was to observe and explore the Mayan and other civilizations.

The first hunters and gatherers settled in the Mayan highlands and lowlands about 11,000 B.C., according to a standard view of archaeologists. But astonishing evidence has been unearthed recently, consisting largely of Clovis points for spears and tools, suggesting that migrants arrived several thousand years earlier, perhaps even from Europe rather than Siberia. Furthermore, they may have been less nomadic and more sophisticated than previously believed. The first Americans were members of our own species, *Homo sapiens.* No bones of prehominids such as Neanderthals, Java man, or Peking man have thus far been found in America, nor have remains of higher primates been found in the New World. So the ancients must have migrated from another continent.

Mayan civilizations occupied southern Mexico, all of Guatemala, Belize, El Salvador, and western Honduras. Traces of one of their houses in Belize date back to 2500 B.C., and pottery that old was found at a site known as Cuello. The Olmec civilization flourished from about 2000 to 400 B.C. along the Veracruz gulf coast, lured there by the fertile farmland of silt deposited by the sluggish rivers. Olmec sculptors created the first great art style of Meso-America. Traces of their art have been found from central Mexico to El Salvador. Writing was developed about 700 B.C., and the first known Mayan solar calendar was carved in stone about 400 B.C.

Many of the Olmec ways endured and became part of the Mayan civilization seven centuries later, about 300 A.D. The Olmec-Mayan links included emphasis on the power of political leaders, warfare that humbled the captives, bloodletting sacrifices, and similarities in their gods. Today there are about 7 million descendants of the Maya, speaking twenty-four related languages.

The original metropolis in the Americas was Teotihuacan, founded about 700 B.C. and located thirty miles northeast of today's Mexico City. By around 500 A.D., up to 200,000 people lived there, making it bigger than Rome at that time. The city included more than 2,600 major structures. Construction of its two great pyramids started about 100 B.C. The city was abandoned about 750 A.D. following extensive fires. Remains of the city cover eight square miles and include the Pyramids of the Sun and the Moon.

The Pyramid of the Sun at Teotihuacan.

A mile-long avenue traverses the center of the city, both sides lined with buildings, including many apartments and temple compounds. On the east side, the ruins are dominated by the Pyramid of the Sun, the third largest in the world, surpassed only by the Tepanapa Pyramid at Cholula, Mexico, and the one at Cheops, Egypt. The Pyramid of the Moon, at the north end of the avenue, has a remarkable cluster of buildings. There the Palace of Quetzalpapalotl is thought to have been the home of the supreme priest. At the south end is the Citadel, a seventeen-acre sunken area that includes the Temple of Quetzalcoatl. This feathered serpent was a god worshiped by Aztec, Toltec, and Mayan civilizations. The bird *quetzal* symbolized heaven and spiritual energy, while the serpent *coatl* denoted earth and material forces. This god was believed to guide man to becoming a solar being.

The Teotihuacans built this enormous city without the benefit of metal tools, the wheel, or beasts of burden. The city was a very pleasant place to live. It had courtyards surrounded by temples and apartments, arranged around an open patio. The plaster walls were decorated with brilliant frescoes. The Teotihuacans were superlative muralists and apparently were a very peaceful people. They influenced many other civilizations and traded with them, including the largest Mayan city of Tikal, 850 miles to the southeast, and the Zapotecs at Monte Alban near Oaxaca. They exercised loose control over all of Mesoamerica, which now consists of central Mexico and Honduras.

Mysteries remain about the Teotihuacans. It is not known who they were, where they came from, or where they went. It is hoped that some answers will be revealed in the future.

Ibberson's initial interest had been the Mayas, but before long he also became engrossed with the Zapotec and Mixtec civilizations, both of whom built and occupied Monte Alban at various times. The Monte Alban ruins are in southern Mexico, six miles southwest of Oaxaca on the Tehuantepec Peninsula. This major religious center was built by the Zapotecs around 600 B.C. on a mountain top overlooking the Valley of Oaxaca. By about 300 A.D., it supported 40,000 inhabitants. Some 600 years later, it was taken over by the Mixtecs, who were later conquered by the Aztecs. The buildings were in ruins when the Spanish arrived.

The Great Plaza of Monte Alban is a grassy area 970 feet long and 650 feet wide, with large ceremonial platforms on each of its four sides. All buildings are aligned on an exact north-to-south axis, except for the astronomical observatory. They all are fascinating, but the Temple of the Dancers attracts the most attention. The oldest building on the site, it is named for elaborate human figures carved into stone, figures whose eyes are closed and mouths are open. Originally they were thought to be dancers in an ecstatic ritual. Later they were proposed to be slain or sacrificed captives. Still another speculation is that their dead bodies were used for medical studies. Many of these stone slabs date back to 500 B.C.

Monte Alban ball court.

Glyphs on them that include numerical signs indicate that a calendar and a writing system already were known then.

Throughout the ruins are underground tombs containing carvings, glyphs, paintings, and frescoes. Some tombs may be entered and may contain original artifacts and remains of buried bodies. Along with them was a priceless collection of more than 500 items that have been put on display at the nearby Oaxaca Regional Museum and constitute the most important treasure ever found in America, second only to that of Tutankhamen of Egypt.

The 1,500-year-old Tomb 7 contained the richest treasure of 180 tombs that have been discovered. Originally used by the Zapotecs, it was discovered in 1932 by the famous archaeologist D. Alfonso Caso. A prominent Mixtec chieftain and some of his companions were buried there along with jewelry and artwork, including masks and ornaments of gold, carved jade, objects of amber jet and coral, a pearl the size of a pigeon egg, a rock crystal goblet, and items made of silver, bone, ceramic, and obsidian.

During one of his Mexican trips in the 1960s, Ibberson was fortunate in being able to acquire a twenty-four-carat gold and turquoise replica of a Zapotec jewel from Tomb 7. It was one of several exact duplicates authorized by the Mexican government. Evon wore the brooch often and thus got a lot of attention.

Tikal, located in northern Guatemala about 50 miles from the Mexican border, is the largest and perhaps the oldest of the Mayan cities. It had links with the Teoti-

Ibberson at the Temple of the Dancers beside a carved human figure, speculated to be a body used for medical studies.

huacan civilization 850 miles to the northwest, having similar ceramic designs and the same storm god, Tlaloc.

Tikal had been settled by at least 600 B.C., but most building took place from 250 to 900 A.D. Set in a high canopy jungle, Tikal has at least 3,000 buildings and monuments, including impressive tall temples, pyramids, plazas, and ball courts. Temple IV, built in 741 A.D., stands 212 feet high, the tallest Mayan structure ever built. At least 100,000 Mayas lived in Tikal when the population was at its peak.

Throughout Tikal are large stone stelae on which hieroglyphics are carved. In the 1970s, their full meaning became understood, though research on the details still continues. Stela 29, dated 292 A.D., is the earliest monument discovered in the Mayan lowlands. A stela erected in 869 A.D. is the

Zapotecos brooch acquired by Joe, one of the duplicates of a 1932 find authorized by the Mexican government in 1960, made with twenty-four-karat gold and turquoise.

Above: *Temple IV at Tikal, at 212 feet the tallest Mayan structure.* Right: *Stelae at Tikal with hieroglyphics depicting the history of thirty-three rulers and the Mayan calendar.*

most recent one known in the Mayan world. A wealth of information was recorded on the stelae, including the accurate Mayan calendar and people associated with the royal court. There is a history of thirty-three rulers, including one queen. Others mentioned are nobles, priests, messengers, ball players, singers, trumpeters, dancers, tortilla makers, weavers, artists, and scribes who recorded events and kept account of tribute.

Of all the Mayan ruins, Chichen Itza, located in the north-central part of the Yucatan Peninsula, draws the most visitors. Founded about 435 A.D., it consists of pyramids, temples, and shrines. The Toltecs arrived from the north about 900 A.D. and exerted considerable influence in the city, until it was abandoned about 1200 A.D.

When the Ibbersons visited these ruins during the 1950s and 1960s, they experienced a very isolated, primitive adventure. During one trip, they encountered only two other visitors, a couple from Denmark. Now visitors arrive by the thousands from a nearby airport, cruise ships, or the popular Cancun resorts. Hotel facilities and highways today are excellent.

The pyramid El Castillo, "the Castle," is one of the best preserved in the Mayan world. Its features exemplify the Mayan calendar. Each of the four sides has 91 steps; the total of 364 plus the platform at the top equals the 365 days in the Mayan year. There are 18 terraces on the sides, representing the 18 months in their year. On each side there are 52 carved panels, the number of weeks in their year.

An astronomical observatory built during the tenth century can be climbed by a stairway leading to the dome. Apertures in the dome line up with particular stars on certain dates. Thus the priest could determine the proper timing for critical events, such as when to plant and harvest, how to adjust the calendar, and when gods needed attention.

The Temple of Skulls is decorated with rows of skulls fashioned in the dance of death. They reflect the gruesome act of human sacrifice that was part of the Mayan religious rites. The Temple of the Warriors consists of a three-tiered pyramid surrounded by many columns, called the group of a thousand islands. The walls of a temple remained on the top of the pyramid, but the ceiling had collapsed and disappeared. On the walls were typical Mayan carvings, as well as Toltec warriors and death symbols. A perfect Chac-Mool, a stone altar with a reclining human figure, is centered in the front of the temple.

The Nunnery, so named by the Spanish conquistadors because it reminded them of a convent, is a large complex of buildings with carved designs, animals, and flowers. Chichen Itza also has two ball courts, including one of the largest used by the Mayas. Two teams of players attempted to propel a large rubber ball through stone rings at each end of the court without using their hands. Some believe that the losers were put to death.

Near Chichen Itza is the Sacred Cenote, a 190-foot-wide limestone sinkhole filled with water. There, humans were sacrificed to appease the rain god Choc. Many skeletons have been excavated, including those of young ladies as well as men, women, and children, along with jewelry, idols, and jade artifacts.

The Ibbersons found Uxmal the most beautiful and impressive of all Mayan archaeological sites. About eighty miles southwest of Chichen Itza, it is located in the northwestern part of the Yucatan Peninsula. About 25,000 inhabitants flourished there from 600 to 1000 A.D., but their number then dwindled and they abandoned it completely by the fourteenth century. The structures are completely Mayan, with no evidence of any Toltec influence such as occurred at Chichen Itza. The cut stone exterior walls of buildings were assembled in close-fitting geometric patterns. Amazingly, the buildings were constructed without using metal tools, the wheel, or draft animals. The architecture has ornamental stone facades with intricate designs. Common carvings are serpents, latticework, and masks of the rain god Choc.

The Pyramid of the Magician, 125 feet in height, is the most dominant structure. The pyramid and the temple on its summit consist of five separate layers corresponding to as many periods of construction. This pyramid probably functioned as a ceremonial building where rulers were crowned. Climbing the pyramid was risky, but that did not deter Ibberson from climbing to the top on two separate visits.

The observatory at Chichen Itza has apertures that enabled priests to determine the timing of critical astronomical events.

Above: *Columns surrounding the Temple of the Warriors at Chichen Itza, and the pyramid El Castillo to the right.* Left: *The Nunnery at Chichen Itza, with carved designs of animals and flowers.*

Above: *The Governor's Palace, thought to be Uxmal's administrative center. The fertility stone in front originally was upright.* Left: *Intricate stone facing on the Governor's Palace.*

The Great Pyramid has been restored only partially. The palace on top is decorated with Choc masks and bird carvings. Below it are nine terraced levels.

The Governor's Palace is reputed to be the finest architectural achievement of the Mayans. The 300-foot-long palace looms above three terraces, giving it an overpowering massive appearance. Intricately carved stone figures and geometric designs cover the upper facade. Serpents, masks, and mosaic patterns all blend together. A seated godlike figure in the center has a plumed headdress. On one side of the palace are 103 stone carvings of the rain god Choc, which together form the image of a crawling snake. The Governor's Palace probably was Uxmal's administrative center and also served astrological purposes.

The Spaniards named four buildings in a quadrangle the Nunnery, as the seventy-four small rooms reminded them of nuns' quarters. Each of the four buildings has a unique, ornate facade with beautiful, decorative details—stone masks of Choc, entwined serpents, mosaic patterns, and latticework designs. Archaeologists surmised that the Nunnery was used by Uxmal's elite ruling class.

The House of the Turtles has a border of turtles carved along its upper molding. The turtles were presumed to suffer with man, and therefore would pray to Choc for rain.

A people called Itza began building the city of Mayapan about 1244 A.D. in northwestern Yucatan, about fifty miles from Chichen Itza. They came from Chichen Itza, which they had settled about twenty years earlier after the Toltecs and Mayas had abandoned the city. Uxmal was declining at about the same time.

Mayapan was one of a few walled Mayan cities and had a population of about 12,000 at its peak. There appears to have been no plan for its layout. Broken pyramids and buildings have since become overgrown by the jungle, and when the Ibberson's visited, restoration was proceeding very slowly.

The Itzas abandoned Mayapan about 1461, perhaps signaling the demise of the Mayan civilization. Hernán Cortés met the Itza people in 1524, the last of the Mayas to remain unconquered. By 1541, the Spaniards put a temporary end to Mayan resistance and established a capital at Merida in the Yucatan. In 1821, Mexico became independent of Spain.

Ibberson also was fascinated by the Aztecs, who were a nomadic, primitive, warlike tribe when they arrived in the Valley of Mexico about 1200 A.D. They established a rich, tribute-paying empire that included most of the present-day Mexico and absorbed the culture of nearby civilizations. Their capital city, Tenochtitlan, was built on an island in Lake Texcoco about 1325. By the time the Spanish arrived in 1519, the powerful Aztecs controlled all of central Mexico and as far south as Guatemala, including the Mixtecs and the Zapotecs. They made human sacrifices of their prisoners of war, ripping out their hearts to please the god of war and the sun. This bloody ritual led the Spanish to destroy the Aztec capital and use its stones to build Mexico City.

The arrival of Cortés in 1519 was one of the most incredible coincidences in history. That was the exact year when the Aztec ceremonial calendar prophesied that Quetzalcoatl, the god of all civilization, would return with fair skin and riding on godlike creatures. Upon his arrival with sixteen horses and 400 soldiers, Cortés was showered with gifts by the Aztecs, who believed he was their returning god.

The Aztecs had many calendars based on the sun, moon, Venus, and various stars. To decipher the calendars, one must consider their rings, segments, symbols, numbers, and cycles, especially the Venus cycle of 243 years. The priests who kept the calendars had a very complex, demanding profession. The signs and symbols they had to interpret came from knowledge accumulated over thousands of years from other civilizations, principally the Mayas and Toltecs. The many calendars had to be read together for religious, agricultural, and military purposes.

The solar calendar consisted of eighteen months of twenty days each, followed by five "hollow days" when everyone should stay indoors without working or playing, which would bring bad luck. The solar calendar was important, as it determined when crops should be planted as well as other significant events. The ceremonial calendar kept track of religious life and signs that ruled over business, fertility, and war. It had twenty months of thirteen days each, designated by symbols such as an eagle, flower, or reeds. Each year of the solar calendar was named for the symbol of the ceremonial calendar's day when it started. Cortés arrived in the reed year, as had been prophesied.

APPENDIX C

The Alpha Tree Farm —Joseph E. Ibberson

The Alpha Tree Farm is two miles long and covers 450 acres made up of twenty parcels, most of them acquired from 1960 to 1970. This is really a working tree farm.

Geology. During 300–220 million years ago for a period of 80 million years, the North American and the Ancestral African tectonic plates collided, creating the 2,000 mile long Appalachian Mountains typified in Pennsylvania by the ridges and valleys folded like wrinkles in a carpet. Oceanic fossils, mostly small clamlike brachiopods, are present on this tree farm.

Soils. Laidig-Buchanan-Andour (70%). Deep well drained reddish yellow subsoil. Mostly very stony. DeKalb-Lehew (10%) moderately deep, well drained, chanery sandy loam. Tulip poplar thrives on this soil. Calvin-LecKill-Klinesville (5%) deep to shallow, well drained—shallow silty subsoil—little clay—red soil.

Forest types are not distinct. Mixed oaks are often intermingled with northern hardwoods such as basswood and an occasional beech and sugar maple. There are small pure stands of tulip poplar, white ash, and black birch. White pine and hemlock are rather rare, perhaps due to early fire history.

Conditions when acquired. Several of the parcels were mature prime timber. Other parcels were young pole stands being overrun with vines. About 200 acres urgently needed timber stand improvement, including the cutting of vines. At least 100 acres needed improvement cutting harvests. Some stands were over-mature and harvesting was in order.

Roads. Since there were no existing roads, the first project was to build roads for fire protection, timber harvesting, access for timber stand improvement, and recreation activities. There are now eleven miles of roads, most of them suitable year round for logging trucks.

Timber stand improvement was started immediately and the 100 acres were completed by 1970. Along with the vines, undesirable species and low quality trees were cut. These stands were converted as quickly as possible to high quality, high value species. Now there is quality timber being harvested on these acres.

Improvement cutting harvests were accomplished with three timber sales. On the three rights-of-way that cross the property, only selective basal spraying of tree species is permitted. This results in a dense growth of shrubs and other low-growing plants that provide food and cover for wildlife.

Tree planting. Five thousand white pine, hemlock, and Norway spruce were planted, which the deer greatly appreciated and almost completely harvested soon after planting.

Current harvesting is on a continuous basis. Mature, storm-damaged, dead trees and fuelwood are harvested. Group selection, single tree selection, and salvage cutting would best describe the harvesting methods. Prior to 1994–95 the tree farm had an open parklike appearance. During the summer of '94 and '95 there was extensive defoliation by gypsy moth caterpillars and fall canker worms followed by severe droughts causing significant mortality. The defoliations and mortality resulted in considerable undergrowth, which is desirable for wildlife. All the sawtimber mortality was harvested. The smaller dead trees are gradually being harvested for fuelwood.

Ponds were constructed at four locations for fish and wildlife. Largemouth bass thrive in all four ponds. It is enjoyable to feed the fish with floating pellets. Bald eagles, ospreys, and kingfishers do most of the fishing. They are frequently seen diving for fish.

Wildlife abounds on the Alpha Tree Farm. Many large openings resulting from the timber harvesting provide excellent cover and browse. There are many trees of mast-producing age with adequate space for maximum mast production. Deer herds are often seen during fall and winter with their heads down eating acorns. It is not unusual to see 100 or more deer during a late afternoon safari over the eleven miles of road. Grass openings are maintained throughout the tree farm to provide insects for young turkeys and other bird life. Flocks of turkeys scratching for mast, insects, and roots are quite common. Large flocks of over 100 have been observed. While driving the roads, grouse are frequently seen. Three different bears are known to visit the area. There is a bear den at the eastern end of the tree farm. Numerous coyotes have been seen and heard throughout the year. Other wildlife seen include red and gray foxes, mink, bobcats, red and gray squirrels, rabbits, muskrats, raccoons, and a mountain lion that may have been released. An adequate number of den trees have been left throughout the tree farm. The four ponds attract a variety of ducks, geese, and waterfowl. During the migration season many species of hawks are observed. Great horned owls are frequently seen and heard. Great blue herons are also frequent visitors. Song birds, especially summer migrants, are often abundant. Pileated woodpeckers are numerous.

Wild flower species are numerous. Among them are yellow violets, cardinal flowers, and pink lady's slippers.

Recreation is a major use. Early evening safaris over the eleven miles of road are exciting. Many of the species mentioned under wildlife can be observed in their natural state. Afternoon picnics are often held at the ponds. As evening approaches, everyone remains quiet and observes the wildlife as it arrives to drink, hunt, graze, and play in the grass openings. The ponds are great wildlife magnets, especially on hot summer evenings. Feeding the fish is a lot more fun than fishing for many of us. A vista has been opened, creating a twenty-six-mile breathtaking view of the mountains and valleys and the Susquehanna River. Hiking, jogging, and mountain biking are encouraged. Parts of the tree farm are leased for hunting purposes. The owner and invited friends hunt for deer and turkey.

Education and demonstration values have been impressive. Numerous prospective tree farmers have been given a tour of the Alpha Tree Farm. Most were so impressed they wanted to duplicate it. Some have even done better. Large groups frequently use the area for tours, up to 120 people. The VIPs of the Stewardship program visited with about 100 VIPs. Numerous garden clubs requested and were given tours of the tree farm. During September of 1998 about 100 members of the Pennsylvania Forestry Association participated in a tour of this tree farm. The education and demonstration efforts are continuous and on-going.

General. Tree farming provides health, wealth, and recreation. It can be your gymnasium, chapel, and money machine. It is a way of life. It is good for the soul to get out and enjoy what God has created. While cultivating the best possible trees, one is also cultivating the best possible character and habits. Take care of the land and it will take care of you. This Alpha Tree Farm could no doubt provide a modest income for a family for the foreseeable future. Tree farmers leave a legacy for those who follow in their footsteps.

Credits. The Bureau of Forestry, especially Paul Troutman, has been most helpful in forest management planning and other activities. Ben Mentzer, a logger with experience at the Michaux State Forest and the Forestry School at Mont Alto, has taken a personal interest and is most careful in tree felling and skidding activities. Rick Petroski and his two sons, Patrick and Joseph, enjoy cutting fuelwood and doing general maintenance and operational activities. Bruce Corsnitz, who lives nearby, uses his mowing, spraying, and road-building equipment. He also has taken a personal interest in the properties and is always available to do what is requested.

APPENDIX D

Reminiscences of the Petroski Family

Joseph Ibberson has had a close, enduring relationship with the family of Rick Petroski. Rick and his wife, Betty, are both teachers, and they have two sons, Patrick and Joey. Rick and his sons have helped Joseph develop and maintain his tree farms, which have become a spectacular wonderland. Joseph and Rick regard each other as family members and continually inspire one another with ideas for improving the tree farms. Through this relationship, Joey was motivated to become a forester.

Reminiscences of the Petroski family help in understanding Joseph's personality and the influence he has had on other people. The family has known Joe Ibberson for nearly thirty years through their work together at the Alpha Tree Farm. Upon hearing about the Ibberson biography, they recorded their recollections and opinions of Joe, which have been edited slightly.

Joseph met Rick in the late 1970s, when he hired Rick's father, a concrete contractor. The job to replace sidewalks at Joseph's house went to Rick. The two immediately hit it off as they discussed hunting and conservation. Joseph told Rick about some properties he owned in Fishing Creek Valley and on Peter's Mountain. He invited Rick to cut some wood there. The location was ideal, as the Petroskis live nearby outside of Dauphin, and so their long-lasting relationship began.

Rick couldn't get enough of the tree farms, especially the Alpha Tree Farm. There were over twelve miles of road, so getting firewood was easy. Wildlife was everywhere—deer, turkeys, and bears. Rick recalls when he first took his boys, about three and five years old at the time, to the tree farm. Joseph was showing them one of the ponds. As they walked around the pond, the boys delighted in seeing the pattern of ripples in the water as the largemouth bass kept following them. Joseph picked up grasshoppers and threw them in the water, and the fish fought over the meal. The boys instantly fell in love with the place, and from then on they always wanted to go to the mountains with their dad. The Alpha Tree Farm became Patrick and Joey's childhood playground. In every season, new adventures awaited the boys. During the warm months, they caught and released frogs, toads, tadpoles, and salamanders. Closer to hunting season, they dressed in camouflage clothing and pretended to hunt deer. The boys became experts at identifying tracks left by various animals.

The boys' activities became entertainment for Joseph and his wife, Evon. Patrick and Joey amused the couple with their intriguing questions and their imaginative play. Joseph had shown Patrick some rocks containing fossils. It became little Patrick's mission to find more, and he checked every rock he could pick up. Almost every time he met Joseph and Evon in the woods, Patrick had to show them what he thought were fossilized rocks. Unfortunately, there were only a few real fossils in his collection.

One snowy day, Joseph pulled up in his car to an area where Rick was cutting firewood. A short distance away, he noticed little Joey, five years old, dressed in fluorescent orange and holding a pop gun. Joseph said to Rick, "What's going on over here?" Rick said Joey was pretending to hunt deer just like his dad. As Joseph and Rick walked over to Joey, Joseph noticed the little wooden pop gun that Joey was using as his deer rifle. The adults got a big chuckle out of this. Joseph still tells that story about the little boy who was not much more than a toddler, standing in the cold, snowy mountains hunting for deer.

Betty Ann, who is a reading specialist, read to the boys soon after they were born. They acquired strong background knowledge on a variety of topics. After meeting Joseph and visiting the Alpha Tree Farm, however, the boys' interests quickly focused on books about nature and wildlife.

As time went by, Joseph's wife became ill. Rick truly admired the way Joseph dedicated himself to the care of his wife in her last years. She enjoyed the Alpha Tree Farm as much as Joseph, and he ensured that her ailment didn't prevent her from spending time there.

After Evon's death, Joseph concentrated on maintenance and projects at the tree farms. Rick worked with Joseph in the summers, when he had time off as a teacher. There was much work to be done. Joseph had longed to complete two projects at the Alpha Tree Farm: a vista along a formation of rocks and a hand-built spring made of mountain stones.

A cliff with exposed bedrock along the north edge of the property was a perfect location for a vista. A few trees were removed to open a viewing distance of thirty-six miles, a panorama of mountains, valleys, and the Susquehanna River. Within the rocks were deep crevices that could provide snug locations for animals to hibernate. Joseph asked Rick and Joey to crawl as deep as possible into these openings. They removed debris and dirt to create small caves within the rocks. The debris that was removed provided the base for a trail below the ledges. In

Above: *Joey and Patrick Petroski studying the aquatic creatures.* Right: *Joey and Pat Petroski are proud of a buck their father, Rick, shot.* Below right: *One of the many trophy bucks taken by Joseph Petroski on the Alpha Tree Farm.* Below left: *Coyotes often are shot during deer season; Joe and Pat got this one.*

many places, Rick and Joey built walls of stone to support the trail. Joseph placed a picnic table above the rocks. Thus they completed the Bear Den Vista after several months of hard work. Rick felt it rewarding to be part of this project.

Joseph thought that a hand-dug spring built of mountain stone with a small waterfall would add to the already beautiful scenery around the third pond on the property. An underground spring was one of the many feeders above the pond, where Joseph and Rick had removed undergrowth to highlight huge yellow-poplars and red oaks. The next summer, the two men did most of the digging, and then collected stones to form the walls of the spring. They placed these stones on top of a concrete footer and sealed them with mortar and clay, then installed a pipe to provide access to the spring water. Finally, Rick and the boys moved a huge rock to form a bridge for crossing over the run that feeds the spring. Even though Rick claims the summer heat and humidity almost killed him, he is most proud of this project.

Other projects included the addition of a fourth pond, development of food plots for wildlife, and the construction of new roads. Patrick and Joey were teenagers by then, so they helped with much of the work on the property.

It was during this period that Rick really got to know Joseph Ibberson. Working together almost daily during the summer months, they had many interesting conversations, discussions, debates, agreements, and disagreements. Rick noted that Joseph always gave his full attention when speaking with others. When it came to making an idea a reality, Joseph was always open to others' ideas and opinions. He asked questions, visualized the results, determined whether the idea was worthy, revised as necessary, and finally decided on a plan.

Over the years, it became apparent to Rick that Joseph is a believer in the importance of exercise and maintaining an active mind. Joseph not only supervised, but worked right along with his helpers. Maintaining the tree farms required a lot of physical work, but that didn't deter Joseph from doing his share. He especially enjoyed demonstrating his expertise with the chain saw. The Petroskis always marveled at how this "older fellow" worked right along with them, especially on hot summer days.

To complement the good life Ibberson has lived, he met a really special person. Nila Grimes came into his life several years after the death of his wife. More than a close friend, she has become a helpful partner in his daily activities and his charitable causes. On most days, Nila cooks dinner and they enjoy each other's company in the evening. This has filled a void in both of their lives. It is admirable to observe the respect and caring that they show toward one another.

As of this writing, Joseph is eighty-nine years old. Even though he doesn't do as much physical labor as he used to on his properties, he is still very active at home. He weeds his vegetable and flower gardens and maintains his fruit trees.

Joseph has always believed that daily exercise is an important aspect of having a quality life. He also plants flowers and maintains the home of his close friend Nila.

An active mind is as important as an active body to Joseph, who spends many hours daily in his den reading, researching, and planning. He is quite the expert on nutritional foods and has an impressive collection of resources on health and nutrition. His bookshelves are also filled, as expected, with an extensive array of books and journals on forestry.

Joseph stays current with the ever-changing developments in forestry, and he is still planning improvements to the Alpha Tree Farm. Rick and Joey have laid out corridors for the addition of more roads, and Joseph is considering an improved entrance to the tree farm, as well as a storage shed to house the supplies needed for maintenance. Age has not slowed him down.

The Petroski's son Joey is now a forester. Joey credits Joseph for greatly influencing his career choice. Joseph's tree farms provided Joey with a real education, as he spent countless hours there ever since he was a little boy. Joseph quickly recognized Joey's potential as the teenager worked on the Alpha Tree Farm, and it was he who encouraged the boy to pursue a career in forestry. He repeatedly

Joseph Petroski lecturing Penn State students about the management plan he formulated for the Alpha Tree Farm.

reminded Joey that mathematics is very important and persuaded him to take all of the math courses offered in high school and to become proficient in them. Joey's love of nature made the career decision an easy one. Rick, however, was concerned that it might be difficult to find a job in forestry because the turnover rate was so low. Ibberson objected, saying that one should pursue something one loves to do and it will work out. Joey recalled Ibberson telling him, "If you're good at what you do, you won't have to find work . . . the work will find you."

Throughout Joey's collegiate career, Joseph always wanted to know how he was progressing. He quizzed Joey on courses he was taking. He showed interest in what Joey was doing on his summer internships with the Pennsylvania Department of Conservation and Natural Resources. Joey graduated with distinction from Penn State's School of Forest Resources and found employment a few months later with the DCNR Bureau of Forestry. The bureau administrators were so impressed with his work at the district level that they enabled Joey to work toward a master's degree at Penn State, specializing in forest resources planning and operations research. His proficiency in math undoubtedly helped.

The research that Joey conducted for his master's degree is directly related to the forest management plans that Joseph prepared about fifty years ago. In the meantime, most of the young forests have matured, and timber harvests have been under way. Joey's research has helped develop a model geographic information system that will simulate timber management practices and predict the spatial structure of forests into the future. Through interviews of district foresters, he identified elements of decisions affecting timber sale locations and boundaries. The planning of harvest units and the scheduling of harvests can affect economical, societal, and ecological dimensions of managing forestlands. Thus Joey is carrying forward the next steps in forest management that Joseph initiated long ago.

Joseph Ibberson has been a true mentor for the Petroskis. He provided opportunities for them to enjoy one of the most spectacular areas in Pennsylvania. In many ways, he helped Joey become an outstanding forester. And he has shown them the ingredients for healthy living.

APPENDIX E
Recollections of Ben Mentzer

Maurice Benjamin Mentzer's friendship with Joe has endured about as long as that of the Petroskis, and it also developed into a close relationship. When Ben learned about Joe's biography, he decided to write down his thoughts too. The words that follow are mainly Ben's, but condensed and lightly edited.

Joe met Ben through a recommendation by a chain saw dealer. Joe and his workers had been putting in many hours of labor to improve his tree farms, much of it on timber stand improvement practices. He was buying a lot of chain saws from Ralph Laubach, a Stihl dealer in Thompsontown. Upon asking a salesman if he could recommend a careful logger, Joe was given the name of Ben Mentzer.

Ben first worked for Joe in 1978, when gypsy moths had moved into central Pennsylvania and had begun to devastate the oak trees. The insects were denuding one of Joe's small tree farms, a nice forty-acre stand of young oaks in Potato Valley. To salvage the dead and dying trees, Joe and Ben drew up their first timber sale agreement. They have continued to work together ever since, harvesting timber and doing other work on Ibberson's tree farms.

After Ben had completed the logging job on the Potato Valley tract, Joe arranged for him to do some logging for Alexander Boyd on his 1,500-acre tract on Blue Mountain. At that time, firewood was in great demand, and the gypsy moths were still feasting on oak leaves in the vicinity. Ben's agreement with Boyd included the harvesting of dead timber, trees uprooted by wind, and thinning some stands of trees.

Ben's relationship with Joe has pertained mostly to logging and working in the woodlands of these two large landowners, Ibberson and Boyd. Ben did timber stand improvement, removed trees killed by gypsy moths or damaged by storms, and cut down trees where Joe had plans to build roads, ponds, and vistas. He worked in spans of two or three years at a time to complete a project and do proper cleanup.

Joe sometimes offered Ben advice on personal matters. Once he suggested investing in a particular stock, which Ben purchased at $6 per share and sold just two years later at $60. After Ben and his wife, Jane, had bought a forty-six-acre property next to their residence, they had questions about surveying the property lines. Joe inspected their deed and those of adjoining properties and offered suggestions. Joe also gave sound advice concerning a neighbor who was using Ben's driveway as an access for his used-car business. After checking the neighbor's deed to see that no right-of-way had been granted and contacting the township's zoning officer about constructing a fence and a living border, Ben and Joe composed a friendly letter to the neighbor that informed him of Ben's intentions to install and plant a fence that would beautify both properties and post the driveway as private. In closing, the letter stated, "It is our intent to be fair and just to all our neighbors. We wish the very best for all endeavours, to you, your family, and your business."

Joe and Ben discussed many topics besides their common interests in forestry, such as politics, the economy, religion, and nutrition. Ben has been interested in Bible stories from a very young age, so the two would converse about biblical topics at length. Joe mentioned that some of his favorite scriptural passages were the writings of Saint Paul. Joe considered the story in the book of Genesis about Joseph, who had a brother named Benjamin, as symbolic of himself and Ben as brothers.

Nutrition is another subject the two aging men discussed. Each had quite a bit to say about their diets, anywhere from oatmeal cookies to avocados; carrot, grape, and pomegranate juices; blueberries, cherries, and pears; nuts, meats, tomatoes, and salad greens. Over the years they would compare notes on ideas and concepts gathered from the various books they read on health topics.

Joe and Ben also would discuss gardening, particularly tomatoes. They would phone one another to see which one had the first ripe tomato, and usually it was Joe. Ben always had the largest plants, about five feet tall. But 2005 was an exceptional year, when Ben's four plants grew to be ten feet and bore fruit through the end of September. Joe generously has shared his fresh produce with Ben and has also given him seeds of kale and a hardy variety of parsley to plant. Once Ben was pleasantly surprised when Joe ordered two varieties of pear trees for Ben's newly enlarged property.

Ben has learned many things from Joe and is currently applying those principles to managing his own property. Although he is now seventy-six, he is still cutting timber for Joe on the tree farm and his own land, clearing the underbrush and vines to keep the better tree species growing.

Ben remembers a time when he and his wife were invited to meet Joe and Evon at the picnic table by the lake. It was a beautiful day, and the setting was serene as they all enjoyed good food and conversation while the sun was high in the sky. The beauty of those woodlands can be compared to what the Garden of Eden may have been like.

Joe has done a splendid job as a developer and caretaker of these beautiful woodlands of the Alpha Tree Farm, and we applaud him for all his achievements.

APPENDIX F

*Health and Nutrition**

One does not have to be around Ibberson for very long to realize that he is very knowledgeable about health and nutrition, and many of his friends seek and follow his advice and recommendations. With his inquisitive mind and background in research, he has been a student of nutrition for several decades. He accepts recommendations only after doing thorough research that shows positive, well-documented results. Among his sources of information are best-selling nutritional books and health letters from the Mayo Clinic, Johns Hopkins University, and University of California at Berkeley. *Science News,* a weekly magazine, often is the first to report new nutritional research results, and the monthly *Scientific American* magazine frequently deals with such research in great detail. One of the sources of health and nutritional information he finds most reliable is the *Wall Street Journal*'s Personal Section.

Ibberson's diet takes advantage of his broad nutritional knowledge, and it ties in with his love of gardening.

Phytochemicals, complex chemicals found in plants, have a profound impact on health, and Ibberson believes that it's important to understand their role. Hundreds of these chemicals interact in our bodies with the basic carbohydrates, proteins, and fats, along with vitamins and minerals. These combinations cannot be duplicated in dietary food products or processed foods.

All the beneficial phytochemicals have been discussed in other books, but a consideration of one example will illustrate some principles and the complexity involved. Lycopene, a pigment that provides the red coloration of tomatoes, watermelons, and pink grapefruits, is a powerful antioxidant that helps prevent clogged arteries and heart disease. People who consume large quantities are less likely to die from many forms of cancer, especially prostate cancer in men. Lycopene enhances the ability of skin to withstand damaging rays of the sun. By

*Joseph Ibberson complied this information, which he uses for his diet and good health, believing that it may interest others. His summary illustrates that individuals not trained in nutrition or medicine can keep abreast of the latest recommendations from authoritative sources and apply that information to maintain a healthy lifestyle.

preventing lutein oxidation, it may also lower the risk of macular degeneration and thus protect eyesight. One study showed that elderly people who had high lycopene concentrations in their blood were better able to care for themselves and carry out everyday tasks. Cooked tomatoes provide about five times as much lycopene as raw tomatoes. This may seem counterintuitive, but cooking breaks down the cell walls and liberates the lycopene. Stewed tomatoes, tomato paste, tomato sauce, and ketchup all are high in available lycopene.

Healthful benefits that can result from eating the foods and following the diets recommended by Ibberson include the following:

- Lower risk of heart disease and clogging of arteries.
- An increase in good cholesterol (HDL) and reduction in bad cholesterol (LDL).
- Reduced risk of hypertension, high blood pressure, strokes, and heart attacks.
- Prevention of many types of cancer.
- Prevention of vision impairment, such as cataracts and macular degeneration.
- Prevention of memory loss, dementia, and Alzheimer's disease.
- Prevention of diabetes and osteoporosis.
- Stimulation of the immune system and control of some types of inflammation, thus strengthening defenses against diseases and infections.
- Stabilization of blood sugar.
- Reduction of obesity.
- Decrease in tooth cavities, gum diseases, and dental plaque, and strengthened tooth enamel.
- Relief of depression and reduction of stress.

Many books have been written on what foods to eat and their particular benefits for optimal health. Those listed below, in alphabetical order, have been thoroughly researched by Ibberson. He eats them himself on a regular basis, and assures that they contribute to good health.

- Apricots. Dried apricots are a concentrated source of beta-carotene, niacin, potassium, calcium, iron, vitamin C, and dietary fiber. NASA officials made apricots a key part of the diet for the Apollo moon mission.
- Avocados. Avocados are very high in monounsaturated fat, which lowers bad cholesterol (LDL) levels, and high in beta-carotene and calcium. Their

fat makes soluble vitamins and phytochemicals available to the brain to help prevent memory loss, dementia, and Alzheimer's disease. Carotenoid absorption is increased up to ten times by the fat in avocados.

- Barley. Barley is the best source of beta-glucan, a soluble fiber known to lower LDL (bad) cholesterol as well as blood pressure. It also contain pectin, another soluble fiber. The FDA has ruled that barley products may claim to reduce the risk of heart disease. Barley also is a good source of B vitamins, iron, zinc, and the antioxidant selenium. Using barley in soups is a tasty way to get its benefits.
- Beans, peas, lentils. Colored beans are highest in phytochemicals and are very good sources of low-fat protein, soluble fiber, and water-soluble vitamins. A super heart-healthy food, they lower bad cholesterol (LDL), may help prevent various types of cancer, and play an important role in weight management.
- Blueberries. Blueberries contain vitamins C and E, and antioxidants and phytochemicals that lower cholesterol and protect against heart attacks and urinary infections. They may also help prevent Alzheimer's disease. Eat them daily, fresh or frozen.
- Broccoli. Broccoli is nutrient dense, with cancer-fighting phytochemicals. Eat it fresh or frozen.
- Cherries. Sweet, dark cherries, fresh or frozen, are rich in pectin, a soluble fiber that lowers cholesterol. The dark pigment is due to an abundance of phytochemicals that are healthful antioxidants. They also have anti-inflammatory value and are rich in vitamin C and potassium.
- Cocoa. Cocoa is very high in beneficial antioxidants and contains copper, iron, and zinc. It can be consumed in drinks, candy, and pastries.
- Dates. Dates are high in potassium, iron, and dietary fiber.
- Grape juice, red or purple. Grape juice is a concentrated source of flavonoid phytochemicals that keep bad cholesterol (LDL) from sticking to artery walls. It also contains vitamin C and potassium.
- Kale, spinach. Kale and spinach are top antioxidant sources with amazing health benefits. Be sure to eat a cupful every day.
- Margarinelike spreads. Spreads with added stanols and no hyrogenated oil may reduce bad cholesterol (LDL) up to 15 percent. They are healthful substitutes for butter, regular margarine, or lard.
- Milk. Joe recommends skim milk, which is loaded with high-quality protein, bone-building calcium, riboflavin, potassium, amino acid lysine, vitamins B-12, A, and D.

- Nuts. Nuts, especially raw walnuts, pecans, and dry-roasted almonds, are rich in nutrients, vitamins, and minerals. Have a handful each day.
- Oats. Oats are high in soluble fiber, complex carbohydrates, and minerals. They reduce bad cholesterol, so load up on oats daily in the form of whole-grain cereal, granola, or energy bars.
- Olive oil. The extra-virgin grade oil pressed from olives is a nutritional blockbuster with many biologically healthful compounds. It contains oleocanthol, which provides anti-inflammatory benefits, and is very high in monosaturated fat, which lowers bad cholesterol (LDL). It contains two phytochemicals, hydroxytyrosol and oleuropein, which have antioxidant properties that may protect against clogged arteries, high blood pressure, and cancer. It is also high in vitamin E.
- Oranges. The fresh fruit or orange juice provides calcium and vitamin C.
- Pomegranates. The 100 percent juice of pomegranates is a grand slam, having three times the antioxidant power of green tea or red wine, and also exceeding that of blueberries and cranberries. Drinking the juice daily reduces blood pressure and keeps bad cholesterol (LDL) from clogging arteries. Polyphenols, antioxidants in the juice, suppress free radicals, which can cause aging, heart disease, strokes, and possibly cancer. The juice also is a good source of potassium, B6, and vitamin C. Its ellagic acid is being evaluated for the potential to counteract carcinogenic agents.
- Prunes. Prunes are high in soluble fiber, which lowers blood cholesterol levels. They are a good source of antioxidants, beta-carotene, B vitamins, iron, and potassium. Prunes were at the top of one list of the ten most healthful foods.
- Raisins. Dark raisins are a concentrated source of healthful nutrients including potassium and iron, and large amounts of antioxidants, and soluble and insoluble fiber.
- Salmon. Joe prefers wild Alaskan salmon, fresh, frozen, or canned, which is lower in fat and contains healthful fatty acids, protective against heart attacks.
- Soy. Soy is a high-quality protein rich in vitamins, minerals, soluble fiber, and disease-fighting phytonutrients. Consume it daily in the form of unsalted dry-roasted nuts, tofu, cheese, soy milk, or soy protein powder.
- Strawberries. Strawberries are higher in vitamin C than any other berry and are also rich in dietary fiber and potassium.
- Sunflower seeds. Dry roasted and unsalted, sunflower seeds are a nutrient-dense food, an outstanding source of many vitamins and minerals.

- Sweet potatoes. Sweet potatoes are high in fiber and rich in vitamins, minerals, and phytochemicals. They supply large amounts of potassium, calcium, beta-carotene, and vitamins C, E, and B-6.
- Tomatoes. Use cooked or processed tomato products daily, as they are the principal source of lycopene, which protects against cancer and possibly heart disease. Penn State is working on new tomato varieties that will have three times the lycopene content of typical tomatoes.
- Turkey breast. Skinned turkey breast is perhaps the leanest protein source among meats. It also has an abundance of beneficial nutrients, including vitamin B-6, B-12, niacin, selenium, iron, zinc, and riboflavin.
- Wheat bran. Bran is high in fiber and a nutritional storehouse of magnesium, iron, and selenium.
- Wheat germ. The embryo of the wheat grain is packed with nutrition to germinate and start a new wheat plant. Sprinkle wheat germ on cereal or yogurt, or add to cooked or baked dishes.
- Whole-grain foods. Buy only whole-grain foods. Avoid products with refined grains, such as white bread, pasta, white flour, and white rice, which may have a negative health effect. Whole grains lower the risk of strokes, coronary disease, diabetes, obesity, hypertension, osteoporosis, and certain cancers.
- Yogurt. Use plain, nonfat yogurt with at least three live active cultures, such as *L. acidophilus, S. thermophilus,* and *B. bifidus.* These friendly bacteria, known as probiotics, take up residence in your intestines, where they relieve digestive disorders, rev up the immune system, alleviate some allergies, help the digestive system absorb nutrients, and reduce hypertension and cholesterol. They may help protect against colon cancer and urological infections. Added vitamin D aids in calcium metabolism.

Spices can make foods flavorful without adding lots of fat and salt. Ibberson currently is evaluating recent research reports indicating that spices also are loaded with disease-combating substances. Further research is needed, but spices may aid in fighting inflammation and infections, inhibiting cancer-causing enzymes and tumor-stimulating hormones, and slowing the growth of and destroying cancer cells. Ibberson has gathered the following information on several spices:

- Cinnamon. A USDA study found that consuming cinnamon for forty days reduced the blood levels of sugar and triglycerides—artery-clogging fats—by 25 percent and bad cholesterol levels by 20 percent.
- Ginger. In numerous studies, ginger has been shown to relieve swelling and pain from arthritis. It also has a blood-thinning effect.

- Oregano. A USDA study showed oregano to be very high in antioxidants.
- Sage. A British clinical trial found that sage boosted memory alertness and calmness. Sage oil may improve the memory and attentiveness of Alzheimer's patients by inhibiting the breakdown of acetylcholine, which is critical for memory and thinking but is produced less and less as the disease progresses.
- Turmeric. Curcumin, which gives turmeric its yellow-orange color, may inhibit the breakup and accumulation of destructive proteins that cause Alzheimer's disease. Several tests are under way to determine whether curcumin can prevent the buildup of mucus in the digestive systems of cystic fibrosis patients or treat multiple myeloma and pancreatic cancer. Studies have shown that curcumin kills human breast and liver cancer cells and has anti-inflammatory and cholesterol-lowering properties. Tests on mice indicate that curcumin may be useful in preventing Alzheimer's disease and that it stops the proliferation of prostate cancer cells.

The key to health is a balanced, nutritious diet. Each individual can use information about the benefits of various foods to fashion a healthful diet to suit his or her own tastes. The diet should contain plenty of fruits and vegetables, whole grains, fish, and modest amounts of lean meat. The healthful substances in foods interact and carry out their function as part of a nutrient package. Ibberson's recommendations for a healthy diet are as follows:

- Vegetables and fruits. Eat five or more servings daily of vegetables and fruits. Include those that are orange, red, yellow, and green, pigments that make fruits and vegetables extra healthful. Benefits from phytochemicals, especially flavonoids, include fighting inflammation, cardiovascular disease, cancer, elevated blood sugar, and obesity; they prevent memory loss and possibly help overcome a genetic predisposition to Alzheimer's disease. Be sure to include members of the Cruciferae group, including broccoli, brussels sprouts, cabbage, cauliflower, kale, turnips, and rutabagas. They are nutrient-dense vegetables with incredible health benefits, including the fighting of cancer.
- Whole grains. Eat five to seven servings daily of whole-grain foods, including bread, cereals, tortillas, noodles or pasta, rice products, and whole-grain snacks. Whole grains lower the risk of strokes, heart disease, diabetes, certain cancers, hypertension, and osteoporosis. Whole grains have bran, the outer fiber-rich layer containing minerals, B vitamins, protein, and phytochemicals. The middle layer contains healthy carbohydrates, proteins, and B vitamins. The inner layer, the germ, contains

phytochemicals including vitamin E. When grains are refined to make white flour, the bran and germ are stripped away, along with all their beneficial nutrients, leaving a starch substance high in calories. Whole-grain foods also taste better. The synergy of the different nutrients in a variety of grains provides optimum nutrition.

- Proteins. A moderate amount of protein should be consumed each day, in the form of beans, eggs, nuts, fish, and poultry, especially skinless turkey breasts. Fish, such as salmon, should be eaten twice a week.

- Dairy products. Eat two to three servings daily of milk, yogurt, and cheeses. Dairy products are an important source of dietary calcium, which builds strong bones, preventing osteoporosis and brittle bones that fracture easily, and regulates muscle contractions and the heartbeat. Calcium, along with other minerals, controls blood pressure and reduces the risk of strokes. It may lower cholesterol and prevent blood clots.

  Dairy products also provide other necessary vitamins and minerals as well as an ample supply of complete protein, including the amino acid lysine, which is lacking in whole-grain products, making milk an ideal complement. Yogurt is especially rich in calcium, complete protein, B vitamins, and minerals.

- Fiber: Adequate amounts of both water-soluble and insoluble fiber should be consumed daily. Among its multitude of health benefits are the reduction of heart disease, helping manage diabetes by keeping blood glucose in check, preventing colon cancer, and keeping bowel movements normal.

- Fats: Saturated and trans (hydrogenated) fats should be avoided and kept to a minimum. These fats stimulate cholesterol production, which blocks arteries and leads to heart attacks and strokes. However, mono-unsaturated and poly-unsaturated fats, which occur in canola, olive oil, and avocados are beneficial. Recent research has shown that consuming unsaturated fats along with colorful vegetables can greatly increase the absorption of healthful carotenoids such as lycopene and lutein.

- Salt. Sodium chloride can contribute to high blood pressure. It should be limited to no more than the equivalent of one teaspoonful per day. Avoid salty foods by reading the labels, which specify sodium contents.

- Sugar. Sugary foods should not be consumed in significant amounts. They are very high in calories and void of healthy nutrients. Sugary foods like cakes and pies are also high in fat, and they contribute to obesity and diabetes and cause tooth decay.

- Vitamins and minerals. These should come mostly from foods, not supplements. Many compounds in foods work more effectively, in beneficial ways that cannot be duplicated in supplements.
- Cholesterol. Intake of cholesterol should be limited to 300 milligrams or less daily. Cholesterol is present only in animal products such as meat, dairy products, and egg yolks. It clogs arteries and thus can cause strokes and heart attacks. Foods that lower cholesterol include walnuts, pecans, almonds, beans, oats, soy products, olive oil, salmon, citrus fruits, berries, apples, prunes, apricots, carrots, sweet potatoes, and brussels sprouts.
- Fluids: Normal body functions such as perspiration, excretion, and exhaling moisture consume two to three quarts of water per day. To provide adequate fluids and avoid dehydration, drink at least eight eight-ounce glasses of fluids daily. This can be in the form of water, juice, milk, or soups. Some solid foods also may contribute to your fluid intake, such as fruits and vegetables.
- Charred foods. Dangerous cancer-causing chemicals are contained in charred foods, especially meats. Smoke from fat drippings deposit the same cancer-causing chemicals on food being grilled. Processed meats also contain dangerous chemicals when they are prepared with high heat.

The only way to be sure you are consuming the proper amount of healthy nutrients is to read the food labels and use the information to plan a balanced diet. The Food and Drug Administration requires nutritional facts to be printed on all packaged foods. Data must be provided for specific nutrients, including total calories, total fat, saturated fat, cholesterol, sodium, vitamins A and C, calcium, and iron. The manufacturer may elect to provide additional information. The percent daily value of each nutrient in a standard serving size is also listed. The daily recommended intake is for a person who consumes 2,000 calories per day, which you can adjust to your own daily intake. Foods labeled organic must be grown on farms that are certified by state or private agencies accredited by the USDA. Claims of health benefits may appear on food labels if they are scientifically established and approved by the FDA.

Ibberson adheres to his own recommendations for a healthful diet. He practices what he preaches and is not shy about offering nutritional advice to his friends. His own good health is a persuasive argument for healthy living. He has never taken prescription drugs, except antibiotics to avoid infection after dental procedures. At the age of eighty-nine, he looks great, gardens actively, works on his tree farm, and exercises a keen mind in managing his stock portfolio and keeping up with the news. Joe continues to work on philanthropic activities and stays in touch with his many friends, family members, and acquaintances.

BIBLIOGRAPHY

Anon. 1957. Division of Forest Advisory Services. *Penna. Forests* 37(1): 9.

Anon. 1999. Ibberson donates 350-acre woodland, state establishes first Conservation Area. *Penna. Forests* 90(2): 16–17.

Anon. 2001. Ibberson honored. *Penna. Forests* 92(3): 5.

Anon. 2002. 2001 Rothrock Award to Joseph E. Ibberson. *Penna. Forests* 93(2): 3.

Anon. 2004. SAF golden member Joe Ibberson donates first-ever PA township conservation area. *Allegheny News* 13(2): 14.

Anon. 2004. PFA life member donates first township Conservation Area. *Penna. Forests* 95(4): 6.

Brooks, R. O. 1952. Chemical weeding report for 1951. *Penna. Forests and Waters* 4(3): 60–61.

Clepper, H., ed. 1957. *Forestry education in Pennsylvania.* Penn State–Mont Alto Forestry Alumni Assoc., University Park, PA.

Cook, A. E. 1997. *The Cook Forest: An Island in Time.* Falcon Press, Helena, MT.

Craighead, F. C., and C. L. Morris. 1952. Possible importance of insects in transmission of oak wilt. *Penna. Forests and Waters* 4(6): 126–129.

Division of Research. 1951. Oak wilt discovered in Pennsylvania. *Penna. Forests and Waters* 3(1): 8–9.

———. 1951. Japanese honeysuckle eradication. *Penna. Forests and Waters* 4(1): 13, 24.

Glattfelder, C. F. 1951. Management plans for Pennsylvania's state forests. *Penna. Forests and Waters* 3(4): 74–75, 87.

Hadley, B. L., Jr. 1952. Lead arsenate sprays to control white pine weevil. *Penna. Forests and Waters* 4(1): 10–12, 23.

Ibberson, J. E. 1949. A preliminary report on the diseased condition of red pine in southeastern Pennsylvania. *Penna. Forests and Waters* 1(6): 124–125, 141.

———. 1949. Long delayed thinning program being conducted in state forests plantations. *Penna. Forests and Waters* 1(6): 137–140, 144.

———. 1949. Cook Forest Park. *Penna. Forests and Waters* 1(7): 148–151, 160–163.

———. 1949. Chemical plant killers as new forestry tools. *Penna. Forests and Waters* 1(8): 180, 190–191.

———. 1950. Importance of seed source. *Penna. Forests and Waters* 2(1): 8–9.

———. 1950. Changing liabilities to assets. *Penna. Forests and Waters* 2(2): 37.

———. 1950. What is known about the dreaded oak wilt. *Penna. Forests and Waters* 2(4): 76–77, 94–95.

———. 1950. Types of thinnings vs. glaze damage in plantations. *Penna. Forests and Waters* 2(4): 86.

———. 1951. The chestnut story. *Penna. Forests and Waters* 3(2): 32–37.

———. 1951. Cultural measures for timber stands. *Penna. Forests and Waters* 4(4): 78–84.

———. 1955. Everyone to benefit from State Forests. *Penna Forests* 35(2): 32–35.

———. 1955. New diameter increment rule. *J. For.* 53(1): 34.

———. 1963. Japanese study Pennsylvania's conservation program. *Penna. Forests* 53(3): 43.

Ibberson, J. E., and F. E. Egler. 1951. Right-of-way-maintenance by selective application of selective herbicides. *Penna. Forests and Waters* 3(6): 114–115, 125.

Ibberson, J. E., A. B. Mickalitis, J. E. Aughanbaugh, C. L. Morris, and J. M. Francis. 1952. *Common Trees of Pennsylvania.* Penna. Dept. of Forests & Waters, Harrisburg, PA.

Ibberson, J. E., A. B. Mickalitis Sr., and S. Kurtz. 1951. The cedar-of Lebanon. *Penna. Forests and Waters* 3(5): 96–104.

Ibberson, J. E., D. S. Nace, and A. B. Mickalitis. 1959. State forest tree nurseries and the reforestation challenge in Pennsylvania. *Penna. Forests* 49(3): 85–86.

Ibberson, J. E., and J. C. Nelson. 1961. Stepped-up program necessary to get multiple benefits from state forests. *Penna. Forests* 51(1): 117.

———. 1961. Oak wilt—a second look. *Penna. Forests* 51(2): 69–70.

Ibberson, J. E., and H. Streator. 1952. White pine blight responds to fertilizer applications. *Penna. Forests and Waters* 4(2): 30–31, 46.

Johns, H. 2005. Letter to the editor. *J. Arboric.* 31(3): 153.

Lacasse, N. L. 2003. Honors to Ibberson, a legendary figure in Pennsylvania forestry. *Penna. Forests* 94(4): 20–21.

Mickalitis, A. B. 1944. Forest research in Pennsylvania. *Forest Leaves* 34(4 & 5): 9–10.

Mickalitis, A. B., and D. B. Kutz. 1949. Experiments and observations on planting areas "stripped" for coal in Pennsylvania. *Penna. Forests and Waters* 1(3): 62–66.

Morris, C. L. 1951. The Dutch elm disease. *Penna. Forests and Waters* 3(3): 52–55.

———. 1951.The spread of oak wilt and progress of related studies. *Penna. Forests and Waters* 4(2): 33, 47.

Schaller, L. D. 1951. Better seeds for better trees. *Penna. Forests and Waters* 3(4):76–81, 88.

Taylor, J. J. 1952. Oak wilt—location is the payoff. *Penna. Forests and Waters* 4(4): 86–87.

INDEX

ABOUT THE AUTHOR:
Joseph E. Ibberson's Perspective

When I was informed by Norm Lacasse that the Pennsylvania Forestry Association wanted to have my biography written and published, I was hesitant at first. But I agreed to the project when it was pointed out that a main purpose would be to encourage and show others how to participate in beneficial philanthropic and professional activities. A few weeks later, I was sure I had made the right decision when I was informed that Dr. Henry Gerhold had agreed to write the biography. Henry already had written several books and participated in conservation activities at the state and the national levels. He is well known and highly respected throughout the forestry profession.

Henry Gerhold, professor of forest genetics, has been in the School of Forest Resources at Pennsylvania State University since 1956. Born in 1931, the son of German immigrants, he received a B.S. and M.S. at Penn State and a Ph.D. at Yale University. He is best known for his books, *Breeding Pest-Resistant Trees* (1966), *Landscape Tree Factsheets* (2001), and *Our Heritage of Community Trees* (2002), and two chapters in the *Handbook of Urban and Community Forestry in the Northeast* (2006). He produced a large number of scientific and popular articles, many of them jointly authored with his graduate students.

Professor Gerhold started doing Christmas tree research in 1956. For five decades, he has been breeding improved varieties of Scotch pine, Douglas fir, and Fraser fir. The genetically improved varieties have helped keep Pennsylvania one of the nation's top growers of Christmas trees. The breeding program that Gerhold started has recently provided superior trees for propagation by the Schatz Center for Tree Molecular Genetics, directed by molecular geneticist John Carlson. The double helix model of DNA had been discovered in 1953, just three years earlier. Such terms as *genomics, genetic modification,* and *gene sequencing* were unheard of. Tree molecular genetics had not yet arrived. Seed dealers got seed from wild trees as cheaply as possible.

Professor Gerhold has had a positive impact on the conservation of our natural resources. He has set a standard of excellence that continues to inspire and has distinguished himself as a professor with vision, wisdom, and dedication. His quiet generosity and philanthropy add to his outstanding career. His legacy throughout Pennsylvania includes thousands of improved, more valuable Christmas trees, landscape trees that beautify our streets and parks, and many highly trained professionals to continue his heritage.

His five decades of research and publication of results prepared Professor Gerhold well for writing this biography. During the many interviews, he uncovered every last detail and discovered many interesting situations, some of which I had forgotten until my memory was jogged. I often felt as though I were being questioned by an astute professional prosecutor. Information provided from memory was checked against historical records, with other people involved, and on the Internet to avoid having mistakes appear in the biography. Because I also have a background in research, I understood and appreciated Professor Gerhold's methods. The interviews continued for several years, during which time Dr. Charles Strauss, director of the School of Forest Resources, often participated and was most helpful.

Professor Gerhold certainly has accomplished the mission and purpose of this biography: to prepare a blueprint for others to perform good deeds in philanthropy and professional practices.